VOLUME ONE HUNDRED AND TWO

ADVANCES IN GENETICS

Darwin's Pangenesis and Its Rediscovery Part B

ADVANCES IN GENETICS, VOLUME 102

Serial Editor

Dhavendra Kumar, MD, DSc, FRCP, FRCPCH, FACMG
Division of Cancer & Genetics, Heath Campus, Cardiff University,
Cardiff, UK

VOLUME ONE HUNDRED AND TWO

ADVANCES IN GENETICS

Darwin's Pangenesis and Its Rediscovery Part B

Edited by

DHAVENDRA KUMAR, MD, DSc, FRCP, FRCPCH, FACMG
Division of Cancer & Genetics, Heath Campus, Cardiff University, Cardiff, UK

ELSEVIER

ACADEMIC PRESS
An imprint of Elsevier

Academic Press is an imprint of Elsevier
125 London Wall, London EC2Y 5AS, United Kingdom
525 B Street, Suite 1650, San Diego, CA 92101, United States
50 Hampshire Street, 5th Floor, Cambridge, MA 02139, United States
The Boulevard, Langford Lane, Kidlington, Oxford OX5 1GB, United Kingdom

First edition 2018

ISBN: 978-0-12-815129-7
ISSN: 0065-2660

Publisher: Zoe Kruze
Acquisition Editor: Zoe Kruze
Editorial Project Manager: Peter J. Llewellyn
Production Project Manager: Abdulla Sait
Cover Designer: Mark Rogers

Typeset by TNQ Technologies

CONTENTS

AUTHOR'S BIOGRAPHY

Yongsheng Liu received his bachelor's degree in horticulture from Henan Agricultural University and the master's degree in horticulture from Northwest Agriculture and Forestry University. After graduation, he joined Henan Institute of Science and Technology in 1989, where he was appointed director of Horticulture Research Institute in 2002, and has been professor of Henan Collaborative Innovation Center of Modern Biological Breeding and School of Life Science and Technology since 2017. He was also a visiting scholar at Department of Horticulture, Shandong Agricultural University in 1996–97, and is currently at Department of Biochemistry, University of Alberta. Mr Liu was honored as "Academic and Technical leader of Henan Province" in 2000 and "The National Advanced Individual in Ethics of the Teaching Profession" in 2001. He has published over 120 peer-reviewed research articles and reviews in the areas of genetics, horticulture, biological chemistry, and plant breeding. His major current research interest focuses on Darwin's Pangenesis in relation to plant graft hybridization, the inheritance of acquired characters, parental age effects, inherited diseases, transplantation, xenia, telegony, and other genetic phenomena.

PREFACE

This volume is the second part (B) of the two-volume work on Darwin's Pangenesis. Like part A, it also contains five chapters. Chapter 1 compares Darwin's contributions to genetics with Mendel's legacy, proposing that Darwin should have been regarded as one of the most important pioneers in genetics. Chapter 2 reviews Darwin's concept of graft hybridization and provides further evidence in support of it, emphasizing that Darwin's pangenetic explanation for the production of graft hybrids is consistent with the current understanding of the molecular mechanism—horizontal gene transfer. Chapter 3 focuses on Darwin's Pangenesis in relation to inherited diseases, cancer metastasis, regenerative medicine, transplantation, and organ transposition. It is proposed that the notion of Darwinian medicine should be expanded with inclusion of Darwin's Pangenesis. Chapter 4 focuses on Darwin's Pangenesis in relation to xenia, telegony, reversion, bud variation, and the inheritance and non-inheritance of mutilation. Chapter 5 compares Darwinism with Lamarckism, neo-Darwinism (including the Modern Synthesis), and Lysenkoism. The rediscovery of Darwin's Pangenesis sheds new light on Darwin's own synthesis: natural selection and Pangenesis. It is proposed that Darwinism throws more light on genetics and evolution than any other theories, and there is a need to go back to Darwinian Synthesis.

This two-volume work is not written by a professional historian but more from the position of an experimental researcher for whom it is important to fill the gap between historians and molecular biologists. Historians are more aware of the past but less aware of the present. Molecular biologists are more aware of the present and less aware of the past. For example, historians repeatedly described how Galton disproved Darwin's Pangenesis by his blood transfusion experiments, but they seem to be unaware of the new evidence for heritable changes induced by blood transfusion. Molecular biologists are familiar with circulating cell-free DNA, mobiles RNAs, and extracellular vesicles, but they appear not to be aware of Darwin's so-called gemmules. This is probably one of the reasons why Darwin's Pangenesis has been ignored for such long period of time.

In the history of genetics, neglect of certain findings and theories is not uncommon. For example, both Mendel's laws of heredity and McClintock's work on transposable elements had been overlooked for decades. In 1958, the centenary of the first presentation of Darwin's theory of evolution by natural selection, Michie proposed that the day of Pangenesis's return "has dawned." In 1968, the centenary of the publication of Darwin's Pangenesis, Towers made an interesting comment: "A temporarily-eclipsed hypothesis may find itself eventually rediscovered in a safe ecological niche, or even, as with those of Mendel, elevated to the heights. ... Perhaps it is only now, a century later, that it will be vindicated by the molecular biologists." Today, Darwin's assumption that invisible gemmules are the carriers of hereditary characters, they are thrown off from cells, and they diffuse from cell to cell and circulate through the body, has been removed from the position of a provisional hypothesis to that of a substantial theory. I believe that the rediscovery of Darwin's Pangenesis, if and when admitted, would, just as the rediscovery of Mendel's work, have a tremendous impact on genetics, evolution, agriculture, medicine, cell biology, and the history of science.

Many thanks once again to Zhaoji Zhou, Xiangming Wang, Anne McLaren, Ted Steele, Lihuang Zhu, Yi Rao, Qi Chen, and my family for their inspirations, constant encouragements and support. I am also deeply indebted to Dhavendra Kumar, Alina Cleju, and Zoe Kruze for providing me with this opportunity to publish this serial book. I apologize to all researchers whose work has not been cited because of space restriction. This book is liable to mistakes or omissions. Suggestions and criticisms for improvements would be greatly appreciated.

YONGSHENG LIU

CHAPTER ONE

Darwin and Mendel: The Historical Connection

Yongsheng Liu[*,§]
*Henan Collaborative Innovation Center of Modern Biological Breeding, Henan Institute of Science and Technology, Xinxiang, China
§Department of Biochemistry, University of Alberta, Edmonton, AB, Canada
E-mail: ysliu63@yahoo.ca.

Contents

Abstract

Darwin carried out a host of carefully controlled cross- and self-pollination experiments in a wide variety of plants, and made a significant and imperishable contribution to the knowledge of hybridization. He not only clearly described the phenomenon of what he called prepotency and what we now call dominance or Mendelian inheritance, but also explained it by his Pangenesis. Recent discovery of small RNAs acting as dominance modifiers supports his Pangenesis regarding the control of prepotency by gemmules. Historical studies show that there is striking evidence that Mendel read Darwin's

Advances in Genetics, Volume 102
ISSN 0065-2660
https://doi.org/10.1016/bs.adgen.2018.05.006

The Origin of Species, which had influenced his paper presented in 1865 and published in 1866. Although Mendel's paper has been considered a classic in the history of genetics, it generated much controversy since its rediscovery. Mendel's position as the father of genetics is being seriously challenged. Darwin's main contribution to genetics was the collection of a tremendous amount of genetic data, and the formulation of a comprehensive genetical theory for their explanation. Over the past 150 years, however, Darwin's legacy to genetics, particularly his Pangenesis, has not been considered seriously by most geneticists. It is proposed that Darwin should have been regarded as one of the most important pioneers in genetics.

1. DARWIN'S CONTRIBUTIONS TO HYBRIDIZATION

1.1 Darwin on Fertility and Sterility in Hybrids, and Effects of Cross and Self-Fertilization

Before Darwin's time, researchers had been obsessed by the species problem, which sexual hybridization was supposed to solve. If hybridization succeeded, or generated fertile offspring, it indicated that the parents were "varieties". If the hybridization failed, or its offspring were sterile, it suggested that they were "species". One of the most important questions pertaining to hybridization that interested Darwin was the matter of fertility and sterility in hybrids (Roberts, 1919b). The chapter on hybridism of his *The Origin of Species* was a first attempt to synthesize the information available on the fitness of interspecific hybrids in animal and plants, so that the rules governing the sterility and fertility of hybrids might be inferred. Darwin maintained that the inability of species to cross "is often completely independent of their systematic affinity, that is, of any difference in their structure or constitution, excepting their reproductive systems." He used to comment on the fact that the hybrids produced from reciprocal crosses often differed in fertility, and that while two species might be difficult to cross, there is no strict parallelism between the difficulty of effecting the cross and the degree of sterility of the hybrids. As the result of Darwin's studies concerning sterility to pollen, he could remove the obsession which had long afflicted the study of species and variety question. As he stated, "It can thus be shown that neither sterility nor fertility affords any certain distinction between species and varieties. The evidence from this source graduates away and is doubtful in the same degree as is the evidence derived from other constitutional and structural differences." (Darwin, 1872; Roberts, 1919b).

In his *The Variation of Animals and Plants under Domestication*, five chapters (from 15 to 19) were devoted to hybridization. He made strong statements affirming the efficacy of hybridization in producing new breeds. As he stated, "crossing, like any other change in the conditions of life, seems to be an element, probably a potent one, in causing variability, the offspring of the first generation being generally uniform, but those subsequently produced displaying an almost infinite diversity of character" (Darwin, 1875). He repeatedly discussed the evil effects of inbreeding and the beneficial effects of crosses, believing that it is a great law of nature that crossing is beneficial, whereas long-continued close inbreeding is injurious. Darwin's opinion as to the reason for the good effects of crossing was based on the long prevalent view that, since animals, and hence presumably plants, benefit from changes in their environment, that probably such changes operate to influence the reproductive cells, or that in some way the reproductive cells receive an extra stimulation, which redounds to the benefit of the offspring. Darwin also discussed the influence of domestication on fertility, and various causes of sterility. More importantly, he described the phenomenon of prepotency, which now is known as dominance.

In 1876, Darwin published his *The Effects of Cross and Self-Fertilization in the Vegetable Kingdom,* presenting the results of an experimental investigation performed over ten years (Ghiselin, 2009). He divided the text into 12 chapters, and the first half was devoted to the species and families he examined and the results he obtained. The next three chapters focused on the characters measured, and summarized the weights and heights, constitutional vigour, seed production and the differences between cross and self-fertilized plants. The following chapters focused on the means of fertilization and the habits of insects. In this book, Darwin systematically characterized the phenomenon of hybrid vigour or heterosis, and provided a wealth of information pertaining to flowers and flowering time, pollen and seed, the behavior and impact of insects, and plant breeding systems (Darwin, 1876; Owens & Miller, 2009). The book includes large amount of data that took, for some species, ten years to accumulate, and in total the investigations covered thirty-seven years of work. The species selected for experiments belonged to widely distinct families from a variety of countries, including 69 families, 158 genera and approximately 224 species. He performed a host of carefully controlled cross and self-pollination experiments, collected the resulting seed and examined its colour, size, and weight, and then germinated the seeds and grew on the plants. By this way, he could furnish striking evidence supporting his idea that crossing was beneficial, whereas selfing was detrimental. For example, in the instances of plants of 57 different species, belonging to 52 genera and

30 different families, Darwin performed the most extensive experiment for the purpose of determining different size, between the offspring of cross-fertilized and of close-fertilized plants (Ghiselin, 2009; Owens and Miller, 2009). The total number of the crossed plants amounted to 1101, and of the selfed plants to 1076. Darwin noticed that the plants derived from crosses between different strains of the same species were taller than that derived from crosses within the same strain, and taller in the latter case than in the case of the offspring of self-fertilized plants. The average ratio of 620 crossed to 607 selfed plants in height was as 100: 86. In another experiment, he also demonstrated that the numbers of seed produced from cross pollination of the plants were significantly higher than those from self-pollination: an average of 12 seeds per capsule from self-fertilization, and 80 seeds per capsule from cross-fertilization. As Owens and Miller (2009) correctly pointed out, Darwin was clearly a careful experimenter who made thorough analyses of the data he collected, and the conclusions he drew from his researches are largely valid and still relevant today.

1.2 Darwin's Description of Different Forms of Flowers

In 1877, Darwin published *The Different Forms of Flowers on Plants of the Same Species*, focusing mainly on Heterostyly, a genetic polymorphism in which two or three floral morphs occur in populations (Weller, 2009). The most common type of heterostyly is distyly, in which two floral morphs occur in nearly equal numbers. Heterostylous species typically demonstrate the reciprocal differences in floral morphological character of distylous *Primula veris* L., a species that Darwin studied carefully for many years (Fig. 1; Fig. 1 from Darwin, 1877). The function of heterostyly, in Darwin's opinion, was to promote outcrossing between different individuals, preventing the expression of inbreeding depression. Crosses between different morphs that have stigmas and anthers at the same level were termed legitimate by Darwin because of their high fruit set and seed production. Crosses between stigmas and anthers at different levels usually result in reduced fruit and seed production, and Darwin called these illegitimate (see Fig. 2).

According to Gilmartin (2015), although Darwin was not the first to find floral heteromorphy in *Primula*, he was indeed the first to recognize the importance of the two floral morphs. Since the publication of Darwin's book, innumerable papers have appeared on this subject, and focused on the questions raised by Darwin about the evolution and function of heterostyly. As Weller (2009) pointed out, Darwin's hypothesis that heterostyly promotes cross-pollination between different morphs has been largely confirmed. Interestingly, Li et al. (2016) have demonstrated that

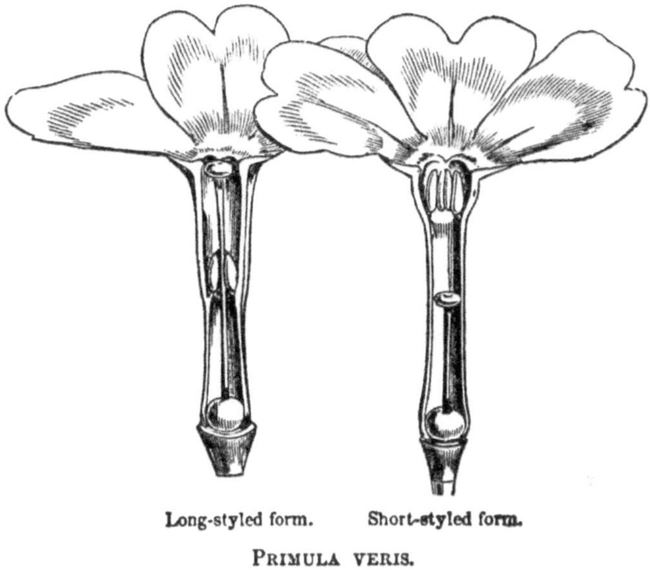

Figure 1 The long- and short-styled floral forms of distylous *Primula veris*. *From Darwin, C. (1877).* The different forms of flowers on plants of the same species. *London: John Murray.*

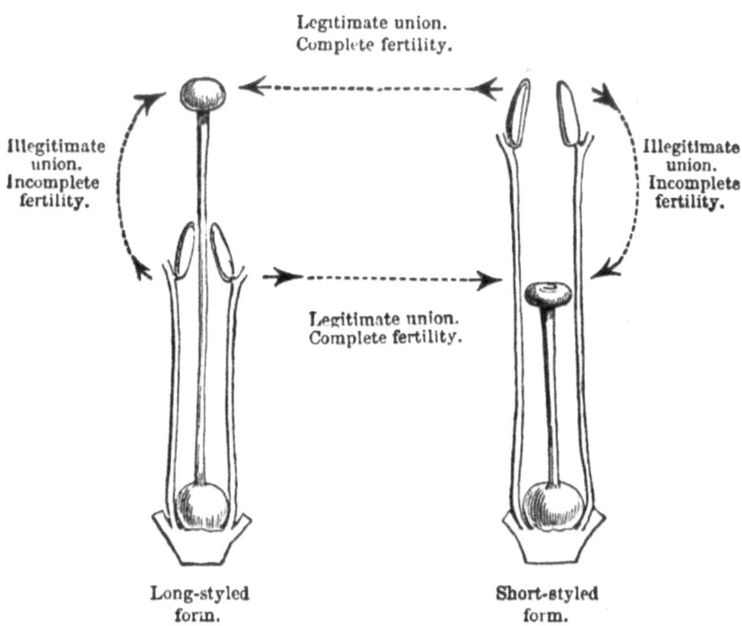

Figure 2 Legitimate and illegitimate crosses in distylous *Primula veris*. *From Darwin, C. (1877).* The different forms of flowers on plants of the same species. *London: John Murray.*

development of the genetically dominant morph in *Primula* is determined by a cluster of five linked genes that is missing in the reciprocal morph. Huu et al. (2016) have showed that an enzyme encoded by one of these genes breaks down a hormone controlling cell elongation, providing an explanation for the reciprocal organ size. Most recently, to test Darwin's hypothesis of disassortative pollination, Jiang, Zhu, and Li (2018) have investigated the pollen sources on each stigma by taking pictures under a microscope, recording the diameter of each conspecific pollen grain with ImageJ. They observed that a shorter distance from the stigmas to the anthers of a pollen donor gave the flower a stronger capacity to receive pollen from the anthers, thus providing further evidence for Darwin's assumption that distyly could promote the pollen transfer between morphs.

1.3 Darwin's View on Yarrell's Law

In genetics, Mendel's laws are well known. Little is known about Yarrell's law. This law was named for William Yarrell, a British naturalist and animal breeder. It maintains that a parent of an older breed will have more influence on the character of the offspring than a parent of a young breed (Liu, 2009). Yarrell's law was of great interest to Darwin, as he was trying to understand why certain breeds seemed to have a greater ability to "impress" their characters on the offspring. To determine the truth of Yarrell's law, Darwin started breeding pigeons. Although the pigeon experiments provided a great deal of counter-evidence to Yarrell's law, he still believed that "Yarrell's law must be partly true", and considered it as one of his favorite notebook laws (Darwin, 1987; Hodge, 1985). According to his Pangenesis, the gemmules of the older varieties were more stable and vigorous, and this was what allowed them to compete successfully against those of the younger varieties (Endersby, 2003).

This law was later confirmed by Ivan Michurin, who found that old varieties of fruit plants possess a stronger capacity for transmitting their characters than young varieties. As Michurin (1949) wrote: "The older the plant chosen as a progenitor, the greater is the force with which it transmits its genes to the offspring, and conversely, if the plant is young, in its first year of bearing, and particularly if it is a hybrid of recent origin, its hereditary power reaches a minimum." Later, it was found that in *Drosophila melanogaster*, the heritability of sternopleural chaeta number was influenced by the age of the parents, those from 14-day-old parent and older are significantly higher than those from 3-day-old parent (Beardmore, Lints, & Al-Baldawi, 1975). Lizana and Sanchez (1994) compared the offspring from crosses between heterozygous individuals of different ages (young × young, young × old, and old × old), and observed an association between adult age and allele frequencies of the offspring. The offspring of the older parents have a greater frequency of the Adf-F allele compared to the offspring of younger parents.

Circulating nucleic acids occur ubiquitously in living organisms. Apoptosis has been considered one main source of circulating nucleic acids. It has been demonstrated that there is an enhanced apoptosis with increasing age (Higami & Shimokawa, 2000; Zhang, Kokkonen, & Roth, 1995). It has also been shown that circulating DNA was significantly associated with age, and increased with increasing age (Chetsanga, Boyd, Peterson, & Rushlow, 1975; Pinti et al., 2014). We may try to explain Yarrell's law by translating Darwin's pangenetic explanation into our modern language. The older parent contains more apoptotic cells, which significantly increases the concentration of circulating nucleic acids. Thus the older parent's circulating nucleic acids have advantage in quantity over those derived from the younger parent. With the establishment that apoptosis is the most common form of cell death and there is an enhanced apoptosis and increased DNA concentration with increasing age, the ability of DNA to be incorporated into gametes and expressed in the progeny, mechanism exists for Yarrell's law (Liu, 2009).

2. DARWIN'S PREPOTENCY AND MENDEL'S DOMINANCE

2.1 Darwin on Prepotency

In the history of genetics, the discovery of dominance has frequently been attributed to Mendel (Casci, 2010). Prior to Mendel, however, this genetic phenomenon had been described by several observers, such as Knight, Goss and Seton (Olby, 1966). Darwin was also aware of the phenomenon of dominance. He described it clearly in term of prepotency, and accounted for it by his Pangenesis.

As early as 1839, Darwin's notebook showed that he was thinking about the question of what he called prepotency and what is now called dominance (Vorzimmer, 1969). In *The Origin of Species*, we read: "When two species are crossed, one has sometimes a prepotent power of impressing its likeness on the hybrid. So I believe it to be with varieties of plants, and with animals, one variety certainly often has this prepotent power over another variety." (Darwin, 1872) In the section "Prepotency in the transmission of character" of his *The Variation of Animals and Plants under Domestication*, Darwin (1875) wrote: "The law of prepotency comes into action when species are crossed, as with races and individuals". He noticed that "in some cases prepotency apparently depends on the same character being present and visible in one of the two breeds which are crossed, and latent or invisible in the other breed; and in this case it is natural that the character which is potentially present in both breed should be prepotent". He concluded that "a character which is present in one form and latent in the other is generally transmitted with prepotent force when the two forms are crossed". Obviously, Darwin's "prepotency" and "latency" are essentially the same as Mendel's "dominance" and "recessivity". Thus Cook (1908) argued that "Darwin was acquainted with a much larger range of Mendelian facts than Mendel himself. Even the Mendelian proportions in the representation of the parental characters were not unknown to Darwin." He stated that Mendel should have read Darwin more carefully and that Darwin understood non-blending inheritance better than Mendel.

2.2 Darwin's Experimental Results Paralleling Mendel's

Darwin not only systematically summarized previous researches on prepotency, but also performed experiments with numerous plant species. For example, there are two types of flower in the snapdragon (*Antirrhinum majus*). The two varieties differ in only one respect: one variety has irregular, radially symmetrical flowers (known as the peloric), and the other has the normal, bilaterally symmetrical flowers. Darwin crossed the peloric with normal plants, whether the cross was of peloric pollen with a normal mother plant or the reciprocal. Among the two beds of seedlings produced, all were normal and not one was peloric. Then Darwin allowed these hybrid offspring (F_1) to self-fertilize, 88 out of the 127 seedlings (F_2) had normal flowers, 37 were perfectly peloric, and only 2 were in an intermediate condition between the peloric and normal state. Today this result is considered as a classic example of dominance, ignoring the two intermediate plants, the proportion of the normal type in the F_2 generation was 0.70, a fair approximation to the Mendelian probability of 0.75.

It is also worth noting that in the table 36 of his *The Different Forms of Flowers on Plants of the Same Species*, it referred to the *Primula auricula* data, where the dominant and recessive forms are cited as occurring in a 3:1 ratio (75 short-styled offspring and 25 long-styled) in crosses between heterozygous short-styled individuals (see Fig. 3). He claimed that he had been

Nature of the Offspring from Illegitimately fertilised Dimorphic Plants.

	—	Number of Long-styled Offspring.	Number of Short-styled Offspring.
Primula veris . .	Long-styled form, fertilised by own-form pollen during five successive generations, produced.	156	6
" " . .	Short-styled form, fertilised by own-form pollen, produced .	5	9
Primula vulgaris .	Long-styled form, fertilised by own-form pollen during two successive generations, produced.	69	0
Primula auricula .	Short-styled form, fertilised by own-form pollen, is said to produce during successive generations offspring in about the following proportions .	25	75

Figure 3 Darwin's table summarizing the results of *Primula* crosses. Note especially the last line referring to the *Primula auricula* data, where the recessive and dominant forms respectively are cited as occurring in a 1:3 ratio in crosses between heterozygous dominant (short-styled) individuals. *Reproduced from Darwin, C. (1877).* The different forms of flowers on plants of the same species. *London: John Murray, and cited in Howard, J. (2009). Why didn't Darwin discover Mendel's laws?* Journal of Biology, 8, *15.*

"informed by a man who raises this species extensively in Scotland about one fourth of the seedlings appear long-styled" (Darwin, 1877). This is consistent with our current knowledge that in *Primula*, the long morph is recessive, and the short morph is dominant (McClure, 2016).

2.3 Darwin's Pangenetic Explanation for Prepotency

The subject of prepotency or dominance can be regarded as a good example of how the same experimental data can be differently explained by researchers committed to distinct theoretical framework. The Mendelian explanation is well known. But little is known about Darwin's pangenetic explanation. According to Darwin's Pangenesis, cells could "throw off" gemmules, which were capable of diffusion from cell to cell or circulation through the body, union with nascent cells and germ cells, and transmission from parent to offspring. He proposed that the tendency of hybrids to resemble one parent might be due to superabundance of gemmules in the fertilized germ, those from one parent having "some advantage in number, affinity, or vigour over those derived from the other parent." In other words, character development depends upon whether gemmules are numerous or few, whether they are in a state of dormancy or not, and whether they have an affinity for other kinds of gemmules. He imagined that the sexual elements of a hybrid included two kinds of gemmules or character-carriers: pure gemmules and combined or hybridized gemmules. He assumed that "when two hybrids pair, the combination of pure gemmules derived from the one hybrid with the pure gemmules of the same parts derived from the other would necessarily lead to complete reversion of character", and "pure gemmules in combination with hybridized gemmules would lead to partial reversion." (Darwin, 1875) A comparison of Darwin's term "pure gemmules" and "hybridized gemmules" with our modern term "homozygote" and "heterozygote" reveals intriguing similarities (Roberts, 1919a).

It is worth mentioning that there are remarkable marginalia by Mendel in Darwin's *The Variation of Animals and Plants under Domestication*, published in German translation in 1868, though it had no influence on Mendel's paper. In the first volume only five entries by Mendel can be found. Mendel paid more attention to the second volume of Darwin's *The Variation*. There are fifty-seven marginalia, mostly in the chapter 27, "On the provisional hypothesis of Pangenesis". Mendel expressed strong disagreement with Darwin's view (Orel, 1996).

2.4 Small RNAs as Dominance Modifier

Ninety years ago, Fisher (1928) proposed the notion of dominance modifier, which is a genetic element specific to an allele that controls dominance-recessivity interactions. For several decades, however, the existence of Fisher's dominance modifiers has been suggested as a theoretical possibility, but these genetic elements have remained elusive (Billiard & Castric, 2011). The self-incompatibility in Brassicaceae is considered as a unique and real example of the dominance modifier model (Goring, 2016). Three genes control self-incompatibility in Brassicaceae, particularly the genes encoding S-locus protein 11 (SP11) on the pollen and S-receptor kinase (SRK) in the stigma. Rejection happens if the parent plants carry the same combination of

SP11 and SRK alleles (called S haplotype). However, the rejection response is also determined by the interaction between dominant and recessive S haplotypes. The question may arise: how does the dominant-recessive relationship between these S haplotypes emerge?

In recent years, small RNAs have been shown to be involved in a broad range of biological processes, and are becoming increasingly recognized as regulatory molecules with the potential to transmit information between cells, organisms and species. In 2010, Tarutani et al. published a landmark paper, showing that a small RNA controlled the dominance of pollen self-incompatibility phenotypes in *Brassica rapa*, epigenetically silencing the expression of the recessive allele. This trans-acting small non-coding RNA is encoded by a genomic sequence that is similar to the recessive SP11 promoters in the flanking region of dominant SP11 alleles. This sequence is specifically expressed in the anther tapetum and processed into 24-nucleotide RNA called SP11 methylation inducer (Smi), which directs the DNA methylation and consequent silencing of the pollen SP11 gene on the recessive haplotype. Billiard and Castric (2011) suggested that these small RNAs correspond exactly to dominance modifiers originally proposed by Fisher (1928).

The intriguing finding by Tarutani et al. (2010) has been confirmed by later researchers. Durand et al. (2014) identified a collection of small RNAs and their respective matching targets that determine the dominance hierarchy among alleles within the gene controlling the pollen S-locus phenotype in *Arabidopsis halleri*. In a recent issue of *Nature Plants*, Yasuda et al. (2016) have identified a new dominance modifier that targets the male SP11/SCR gene to control a dominance hierarchy in *B. rapa* class II S-haplotypes. They demonstrated that a single polymorphic 24-nucleotide small RNA termed SP11 methylation inducer 2 (Smi2) regulate the linear dominance hierarchy of four *SP11* alleles. In all dominant-recessive relationships, small RNA variants from the linked region of dominant SP11 alleles showed high sequence resemblance to the promoter regions of recessive SP11 alleles, acting *in trans* to silence their expression epigenetically. This led them to propose a new model that sequence similarity between polymorphic small RNAs and their target regulates mono-allelic gene expression.

The above recent findings not only provide evidence for the chemical existence of Fisher's dominance modifiers, but also for Darwin's supposed gemmules. Now it has become increasingly clear that living cells release extracellular vesicles, which include small RNAs. The discovery of small RNAs acting as dominance modifiers supports Darwin's Pangenesis regarding the control of prepotency by gemmules.

3. DARWIN' INFLUENCE ON MENDEL

3.1 Earlier Assertion of Darwin's Influence on Mendel

Although there is no evidence that Darwin knew of Mendel's work, there is ample evidence that Mendel read some of Darwin's writings and that those writings had influenced his paper, which published in 1866. It is known that Mendel possessed a copy of the second German edition of *The Origin of Species* published in 1863, many passages of which he marked, and his paper distinctly reflects certain important aspects of Darwin's thought (Bishop, 1996; Fisher, 1936; Fairbanks & Rytting, 2001; Kritsky, 1980; de Beer, 1965). Although Darwin's *The Origin of Species* had no effect on the design or conduct of Mendel's experiments, it may have influenced his interpretation. An interesting point is that in spite of the obvious reflections of Darwin's work in Mendel's paper, he never once mentioned Darwin's name (de Beer, 1965).

In 1936, in his most influential paper "Has Mendel's work been rediscovered?" Fisher noticed the reflection of Darwin's thoughts in Mendel's paper, and proposed that Mendel had read Darwin's *The Origin of Species* before he wrote his paper. In 1964, de Beer further studied this issue in detail, and provided evidence that Mendel owned the German version of the second edition of *The Origin of Species*, published in 1863. The second German edition was based on the third English edition published in 1861. The fact Mendel read the text was established by the 33 pages that Mendel marked. It was Mendel's habit to write page numbers down in the back cover and to mark some pages. A list of the pages marked by Mendel was published by de Beer (1964). From that table it was obvious that Mendel found plenty to interest him in *The Origin of Species*. de Beer (1964) suggested that when Mendel wrote his paper, Darwin's work was very much in his mind. By comparing many passages in Mendel's paper with that of *The Origin of Species*, he concluded that "it is therefore indisputable that Mendel's paper reflects in many passages various aspects of Darwin's thoughts as expressed by him in *The Origin of Species*."

In 1980, Kritsky supported de Beer's conclusion and further revealed several interesting points. He noticed that Mendel cited three pages where Darwin defined creation and evolution. Mendel was particularly concerned with Darwin's comments on hybridization. For example, Page 43, one of the page numbers Mendel wrote on the back cover of the book, included Darwin's discussion about the skill of gardeners. Mendel noted page 111, where Darwin examined the flower and pollination of the pea family. Some pages that Mendel marked concerned Darwin's view about varieties and how they related to species. For example, Mendel marked pages 287 and 303, where Darwin correctly noted that some characters of the parent are prepotent (dominant) in the first generation. Darwin wrote that some individuals "have a remarkable power of impressing their likeness on their hybrid offspring." Darwin had obviously noticed that some characters were dominant. It appeared that Mendel read with interest page 302, since he marked the passage where Darwin dealt with the difference between the first and second generation from a hybrid cross: "The slight degree of variability in hybrids from the first cross or in the first generation, in contrast with their extreme variability in the succeeding generations is a curious fact and deserves attention." Kritsky (1980) exclaimed that Darwin already knew about increased variability in subsequent generations in a qualitative sense five years earlier than Mendel.

3.2 Recent Evidence for Darwin's Influence on Mendel

In 1901, shortly after the rediscovery of Mendel's work, Bateson published an English translation of Mendel's paper, which rapidly became the standard source in English (Mendel, 1901). In recognition of Druery's contribution, it has been referred to as the Druery—Bateson translation (Hartl & Fairbanks, 2007). In 1966, Stern and Sherwood provided a second English version of Mendel's paper, because they felt that the Druery—Bateson translation was seriously flawed (Stern & Sherwood, 1966). This was referred to as the Sherwood—Stern translation, and was generally accepted as the more accurate of the two. According to Fairbanks and Abbott (2016), both the Druery—Bateson and Sherwood—Stern translations share a serious shortcoming: they lack the Darwinian influence that a careful reading and analysis of Mendel's original German article reveal. Thus Fairbanks and Abbott (2016) prepared the third version — a new Darwinized translation of Mendel's paper. This translation has two main objectives. (1). It was intended to be more accurate than the Druery—Bateson translation and more accessible than the Sherwood—Stern translation. (2). It was made an exhaustive effort to use Darwin's phraseology from *The Origin of Species* when choosing English words in the translation.

Under the premise that the passages Mendel marked in *The Origin of Species* might have influenced Mendel's writing, Fairbanks and Abbott (2016) colour-coded words and phrases in Mendel's original German paper that matched words and phrases from the passages Mendel marked in his

German translation of *The Origin of Species*. They then used a different colour to denote words and phrases not in the passages Mendel marked but found elsewhere in *The Origin of Species*. This colour coding indicated that phraseology from the passages Mendel marked in *The Origin of Species* are more frequent and more diverse in the final two sections of his paper, particularly in one paragraph just prior to the final section, subtitled "Concluding Remarks." The following is their English translation of the latter portion of this paragraph. Words corresponding to German words found in Mendel's copy of *The Origin of Species* are in boldface type; those from the passages in *The Origin of Species* marked by Mendel are also underlined. Those labeled with asterisks are cases in which the German-word appeared in *The Origin of Species* but they chose an English-word not used by Darwin:

> No one will **seriously assert** that the **development** of **plants** in a **natural* [free] landscape** is **governed** by **different laws** than in a **garden bed**. Here, just as there, **typical variations** must **appear** if the **conditions of life** are **changed** for a **species**, and it **has** the **ability** to **adapt** to the **new conditions**. It is **freely admitted* [looked]**, through **cultivation** the **production** of **new varieties** is **favoured**, and by the **hand** of **man many** a **variation** is **preserved** that would have **failed** in the **wild* [free] state**, but nothing gives us the **right** to **assume** that the **tendency** for new **varieties** to **form** is so **extremely augmented** that **species** soon **lose** all stability and that their **offspring break up** into an **infinite array** of **highly variable forms**. If the **change** in the **conditions** of **vegetation** were the sole **cause** of **variability**, then one would be justified in **expecting** that those **domesticated plants cultivated** under almost **the same conditions** for **centuries** would have **acquired** stability. As is well **known**, this is not the **case**, for especially among them not only the most **different** but also the most **variable forms** are **found**. Only the **Leguminosae**, like Pisum, Phaseolus, Lens, whose **organs** of **fructification** are **protected** by a keel, **constitute* [make]** an **appreciable exception**. Even for these, **numerous varieties** have arisen during **cultivation** for more than 1000 **years** under the most **diversified conditions**; however, under **the same permanent conditions of life**, they **retain** stability similar to that of **species growing** in the **wild**.
>
> **Fairbanks and Abbott (2016)**

From the new Darwinized translation of Mendel's paper, Fairbanks and Abbott (2016) discovered strong evidence for Darwin's influence on Mendel. They found that a majority of Darwinian terms in this paragraph are from passages Mendel marked. The most obvious is the term *Lebensbedingungen*, which Mendel used twice in his paper. *Lebensbedingungen* appears three times in passages Mendel marked in *The Origin of Species*, in each case corresponding to Darwin's "conditions of life." Fairbanks and Abbott (2016) also noted that a key Darwinian term used by Mendel 10 times, is the

German-word *Element*, which is translatable to English as "element." Mendel used this term to refer to his conception of material hereditary units, which geneticists now refer to as alleles or variants. Although Darwin (1868) began describing underlying hereditary factors as "gemmules" in his Pangenesis, he consistently used the English-word "element" to denote hereditary material in *The Origin of Species*. For example, Darwin's first use of this term in *The Origin of Species* appears in chapter one as "the male and female reproductive elements" near a passage Mendel marked. Fairbanks and Abbott (2016) concluded that their new translation of Mendel's paper provides strong evidence of Darwin's influence, thus supporting Fisher's assertion that in the conclusion of Mendel's paper "the reflection of Darwin's thought is unmistakable".

4. MENDEL'S LEGACY AND LIMITATIONS

4.1 Mendel's Legacy

Mendel's legacy to genetics have been much analyzed and discussed in the scientific literature (Galton, 2018; Hackett, Feldheim, & Alvey, 2006; Liu, 2005; Oldroyd, 1984; Opitz & Bianchi, 2015; Sandler, 2000). Thus it is unnecessary to write too much in this chapter. It should be admitted that Mendel was one of the earlier researchers who applied the knowledge of mathematics and statistics to a biological problem (Dunn, 2003). According to Mayr (1973), "Mendel's outstanding contribution was his experimental methodology, which consisted in the careful counting of all of his observed classes and the calculation of ratios. ... he discovered a new way of looking at inheritance by expressing it in terms of the behavior of unit characters, and he used this interest to arrive at far-reaching generalizations. His achievement was one of the most brilliant in the entire history of science." Mendel's advance over his predecessors, as Fisher (1936) said, was due to crossing closely allied varieties, not different species, which would differ in a large number of different factors. Now Mendel has been recognized as the founder of genetics (Dunn, 2003; Fairbanks & Rytting, 2001; Orel, 1996).

4.2 Geneticists' Criticisms of Mendel's Work

Bateson not only coined the term "genetics", but also was a major exponent of Mendelian genetics. His early enthusiasm for Mendelism was indicated when he named his son Gregory after his birth in 1903 (Davison, 1993).

As he grew older, however, he became more and more convinced that an increasing number of cases of inheritance could not be brought under the Mendelian laws at all (MacBride, 1937). Thus his enthusiasm waned, and he clearly foresaw the failure of Mendelism to explain macroevolutionary change toward the end of his life. In 1924, shortly before his death, Bateson told his son Gregory that it was a mistake to have committed his life to Mendelism, "that it was a blind alley which would not throw any light on the differentiation of species, nor on evolution in general" (Davison, 2004; Koestler, 1971, p. 121). It is interesting to note that Wallace, the co-discoverer with Darwin of natural selection, also denied that Mendelian laws have anything to do with the inheritance of the slight differences that serve as material for natural selection in wild species (Wallace, 1908).

de Vries was a major figure in the history of genetics, and is most widely remembered for rediscovering Mendel's work. However, he considered Darwin one of his most important predecessors because the starting point for his thinking on heredity was Darwin's Pangenesis. Mendelism, in de Vries's later opinion, had nothing to do with "normal" heredity and was mainly a laboratory phenomenon. In 1901, de Vries wrote to Bateson: "I prayed you last time, please don't stop at Mendel. I am now writing the second part of my book [i.e., the *Mutationstheorie*] which treats of crossing, and it becomes more and more clear to me that Mendelism is an exception to the general rule of crossing. It is in no way the rule! It seems to hold only good in derivative cases, such as real variety-characters" (Theunissen, 1994). In 1922, de Vries wrote to his friend Went, regarding an invitation to attend the Mendel celebration in Brno: "To my regret I cannot accede to your request. I just don't understand why the academy would be so interested in the Mendel celebrations. The honoring of Mendel is a matter of fashion, which everyone, also those without much understanding, can share; this fashion is bound to disappear. The celebration in Brno is nationalistic and anti-English, directed especially against Darwin and thus unsympathetic to my mind ...". This suggests that de Vries's short-lived conversion to Mendelism in the spring of the year 1900 did not agree with his previous theoretical framework, thus he had really never been very much of a Mendelian (Meijer, 1985). In a *Science* paper, Cook (1908) appeared to support de Vries's view, and also argued that "Mendelism is not a general phenomenon in nature, nor is it confined to distinct groups of animals or plants, or to particular kinds of characters."

Johannsen was a most distinguished geneticist and the inventor of the term "gene". He also casted a serious doubt as to the evolutionary significance of the Mendelian laws: "To my mind the main question in regard to these units is this: Are the experimentally demonstrated units anything more than expressions for local deviations from the original ("normal") constitutional state in the chromosome? Is the whole of Mendelism perhaps nothing but an establishment of very many chromosomical irregularities, disturbances or diseases of enormously practical and theoretical importance but without deeper value for an understanding of the "normal" constitution of natural biotypes? The problem of species, evolution, does not seem to be approached seriously through Mendelism nor through the related modern experiences in mutations" (Johannsen, 1923).

In his critical paper "Has Mendel's work been rediscovered?" Fisher (1936) found two remarkable facts: (1) the sum of the chi-square statistics of Mendel's different experiments that Mendel gave as an illustration of the variation of that ratio, demonstrates an extremely improbable result. (2) Analyzing the 1:2 ratio of the F2 plants with a dominant trait, Mendel did not consider the loss of recessives. He calculated the expected ratio for an offspring of only 10 plants on the basis of a binomial distribution to be 1:1.7 instead of 1:2. Thus Fisher (1936) concluded that "the data of most, if not all, of experiments have been falsified so as to agree closely with Mendel's expectations". The controversy over the "too good to be true" data remains unresolved and has been the subject of passionate scientific debate (Edwards, 1986; Novitski, 2004a,b; Piegorsch, 1990; Weiling, 1986).

In modern texts of genetics, there are two laws attributed to Mendel. The first is the law of segregation of the factors; the second is the law of independent assortment of genes when several pairs are involved. Monaghan and Corcos (1984) argued that a careful reading of Mendel's paper reveals that neither of these "laws" is to be found in the form now given. Olby (1979) and other historians also held that Mendel was not a Mendelian in any modern senses of term, and that proponents of Mendelism had read much into Mendel's work that was not part of his original formulation (Allen, 2003). Sermonti (1985) regarded Mendelism as a self-producing law. In his opinion, "a law operating only in ideal conditions is no law but a model, a formula, a working tool." If Mendel's "law" of heredity were really a law, then biologists would not worry about how often non-Mendelian inheritance occurs. It has been suggested that non-Mendelian inheritance may be more common than is supposed (Beatty, 1997).

4.3 Breeders' Criticisms of Mendel's Work

Luther Burbank was one of the greatest plant breeders all over the world. In his book *The Harvest of the Years*, he stated that his lifelong adherence to the theory of Darwin was not due to blind faith in Darwin's authority. He tested Darwin's theory in the orchard and field, and the older he grew the more convinced he became that Darwin was the real teacher. He advised a young man who was interested in genetics: "My advice to you is to start Mendel by reading Darwin, and then let Mendel go and read more Darwin". He gave this advice because he saw that the assertions of many well-known scientists were not confirmed and he could do nothing with them practically; whereas he became convinced that Darwin's viewpoints always coincided with the facts (Burbank & Hall, 1927).

Another world famous breeder was Michurin, who was regarded as a "Soviet Burbank". According to his many years' experience in breeding of fruit trees, he argued that Mendel's laws could not be applied in grafted fruit trees. In his *Selected Works*, Michurin (1949) gave a number of facts obtained from the hybridization of fruit plants that indisputably proved the perfect inadequacy of Mendel's law when applied to hybrids of perennial fruiters. He believed that there were many environmental factors influencing the hybridization of fruit plants. Thus he pointed out that "Mendel's conclusions drawn from his experiments on the crossing of two chosen varieties of peas, as well as those of his followers, who effected crosses between different varieties of stinging nettles, barley, maize, etc. could be only accidentally correct; these conclusions would be true only if the same varieties of the above-mentioned plants were used in the crosses and if there were no sharp environmental changes affecting the development of both the parent plants and the hybrid seedlings derived from them." It should be mentioned that, unlike Lysenko, Michurin did not fully deny the significance of Mendel's work. He admitted that "I by no means deny the merits of Mendelian laws. On the contrary, I merely insist on the need to introduce amendments and addenda into it, for it is evident to everybody that his calculations are not applicable to cultivated varieties of fruiters." (Michurin, 1949)

Lysenko was an influential agronomist in the former Soviet Union. He insisted that only science which gave man practical mastery over nature could be true science (Leone, 1952). In the middle 1930s, Lysenko started a campaign, partly political and partly scientific, against Mendelian genetics (Quintanilha, 1959). Now we know that Lysenko was wrong and badly wrong when he mixed science and politics, and regarded Mendelian genetics as "bourgeois science" and "pseudo-science." But some of Lysenko's ideas are thought to be right. As Mather (1942) pointed out, "when Lysenko states that genetics has not contributed very much to the improvements of crops and stocks, we must agree with him. In special cases, like the development of autosexing poultry, genetics has been of great value, but these cases are exceptional. The reason for this failure is not, however, that genetics is unsound, but that its development has, owing to a series of historical accidents, been away from the breeders' problems." Haldane also greatly appreciated Lysenko's practical achievements on the collective farms, and argued that from a practical standpoint, Lysenkoism was a great success (cited in Paul, 1983).

Please excuse me for citing so many unfavorable comments on Mendel. It is not my intention to minimize the importance of Mendel's work in order to exalt Darwin's contributions to genetics, but we must try to see things objectively.

Here I would like to say a few words in Mendel's defence. It is true that Mendel failed to cite Darwin's work. It also might be true that most of his experimental data had been falsified, as Fisher (1936) pointed out. But we must know that he was an amateur scientist, instead of a professional scientist. We cannot make excessive demands on this predecessor.

5. DARWIN, MENDEL AND GENETICS

In 1998, Wang published a book entitled *Explorations into Genetics in Broadened Sense*, in which he argued that Darwin should have been regarded as the true founder of genetics. When I first read it, I felt that Wang must be crazy. But it now appears that there may be an element of truth in his statement.

It should be emphasized that Mendel was not the first to study the phenomenon of heredity, thus was not the first geneticist, as Orel (1996) claimed. Mendel's genius was in depth rather than in breadth, and his place among the greatest of scientists is due to a single paper (Dobzhansky, 1965). He worked at the level of observable characters, and focused narrowly on a problem related to trying to create pure-breeding hybrids and to characterize the identity of species (Allchin, 2003). He was not a great generalizer of masses of heterogeneous data, as Darwin so pre-eminently was. As Russell commented, "For now, Mendelians manifest a certain overconfidence, too much enthusiasm, about the universal application of their ideas ... Mendelism should regard the only appropriate place convenient to itself, which is that of a very precious method of analysis applicable to some modality of hereditary transmission; it could not be considered as a complete solution to all problems of heredity." (cited in Esposito, 2013) Compared to Mendel, Darwin published 25 books and a large number of papers, covering evolutionary biology, genetics, psychology, geology, plant morphology and physiology, and so on. Darwin also published 45 short pieces in *Nature* from 1869, the year *Nature* launched, to 1882, the year of his death. In addition, Darwin wrote more than 15000 letters to colleagues, whereas Mendel wrote only about 10 letters, indicating that Mendel's social life was greatly restricted (Galton, 2018). As far as genetics is concerned, Darwin clearly described almost all genetical phenomena of fundamental importance, such as hybridization, prepotency (dominance), sports (mutation), atavism (reversion), regeneration, heterosis (hybrid vigor), inbreeding degeneration, sex-limited inheritance, inheritance at corresponding ages, graft hybridization, inherited diseases, xenia, telegony, Yarrell's law, the inheritance of acquired characters and many other phenomena (Liu, Zhou, Zhi, Li and Wang, 2009; Muntzing,1959). There was no place for so many facts in Mendel's single paper.

Commenting on Darwin and Mendel, Mawer (2006) wrote: "Had Darwin only read, and understood, Mendel's work, he would have found there the mechanism of inheritance that his theory of natural selection so desperately needed." I fear that this statement is misleading. Historical studies reveal that Mendel simply intended to study hybridization in particular and not heredity in general (Kampourakis, 2013). His interest was the formation of hybrids, not the theory of heredity. In fact, Darwin was the first to formulate a unifying theory of heredity, variation, development, and reproduction. In the late 1800s, it was Darwin's theory, not Mendel's experiments on hybridization, that had the most influence. Darwin's Pangenesis not only greatly influenced many subsequent theories, particularly those of Galton, Weismann, de Vries, Brooks, and Ross, but also is supported by the discoveries of circulating cell-free DNA, mobile RNAs (particularly small RNAs), prions and extracellular vesicles. Mendel's position as the father of genetics is being seriously challenged.

It should be noted that the term "gene", which has passed into everyday language, evolved from Darwin's hypothetical "gemmule" via de Vries' pangen, instead of Mendel's so-called "element". Even the term "element" Mendel used, as Fairbanks and Abbott (2016) pointed out, was frequently used by Darwin in his *The Origin of Species*. Based on these facts, it is not unreasonable to suggest that Darwin should have been regarded as one of the most important pioneers in genetics (Liu & Li, 2016).

In 2008 and 2009, the 150th anniversary of the presentation and publication of Darwin's theory of evolution by natural selection, many papers were published for its celebration (Avise and Ayala, 2009; Beccaloni and Smith, 2008; Costa, 2009; Koonin, 2009; Padian, 2008; Pagel, 2009). Similarly, in 2015 and 2016, the 150th anniversary of the presentation and publication of Mendel's paper, there were also many celebrations (Birchler, 2015; Gayon, 2016; Radick, 2015; Singh, 2015; Smykal et al., 2016; Torres, 2016). 2018 marks the 150th anniversary of Darwin's publication of his big book *The Variation of Animals and Plants under Domestication* and his Pangenesis. Now there is a dead silence without showing any sign of celebration. As Darwin said to Lyell: "My fear has always been that Pangenesis would be a still-born infant, over whom no one would rejoice or cry" (Darwin and Seward, 1903). Lönnig (2017) regarded Mendel's paper as the most famous Sleeping Beauty. But in my opinion, Darwin's Pangenesis is indeed a Sleeping Beauty more than anything else. She has been sleeping for 150 years. However, she is now waking up.

REFERENCES

Allchin, D. (2003). Scientific myth-conceptions. *Science & Education, 87*, 329–351.

Allen, G. E. (2003). Mendel and modern genetics: The legacy for today. *Endeavour, 27*, 63–68.

Avise, J. C., & Ayala, F. J. (2009). In the light of evolution III: Two centuries of Darwin. *Proceedings of the National Academy of Sciences of the United States of America, 106*(suppl.1), 9933–9938.

Beardmore, J. A., Lints, F. A., & Al-Baldawi, A. L. F. (1975). Parental age and heritability of sternopleural chaeta number in *Drosophila melanogaster*. *Heredity, 34*, 71–82.

Beatty, J. (1997). Why do biologists argue like they do? *Philosophy of Science, 64*, S432–S443.

Beccaloni, G. W., & Smith, V. S. (2008). Celebrations for Darwin downplay Wallace's role. *Nature, 45*, 1050.

Billiard, S., & Castric, V. (2011). Evidence for Fisher's dominance theory: How many 'special cases'? *Trends in Genetics, 27*, 441–445.

Birchler, J. A. (2015). Mendel, mechanism, models, marketing, and more. *Cell, 163*, 9–11.

Bishop, B. E. (1996). Mendel's opposition to evolution and to Darwin. *The Journal of Heredity, 87*, 205–213.

Burbank, L., & Hall, W. (1927). *The harvest of the years* (pp. 150–151). Boston and New York: Houghton Mifflin Company.

Casci, T. (2010). Gene expression: Small but dominant RNA. *Nature Reviews Genetics, 11*, 670–671.

Chetsanga, C. J., Boyd, A., Peterson, L., & Rushlow, K. (1975). Single-stranded regions in DNA of old mice. *Nature, 253*, 130–131.

Cook, O. F. (1908). The mendelian inheritance of mutations. *Science, 28*, 86–88.

Costa, J. (2009). The Darwinian Revelation: Tracing the origin and evolution of an idea. *BioScience, 59*, 886–894.

Darwin, C. (1868). *The variation of animals and plants under domestication*. London: John Murray.
Darwin, C. (1872). *On the the origin of species by means of natural selection; or the preservation of favoured races in the struggle for life* (6th ed.). London: John Murray.
Darwin, C. (1875). *The variation of animals and plants under domestication* (2nd ed.). London: John Murray.
Darwin, C. (1876). *The effects of crossing and self-fertilization in the vegetable kingdom*. London: John Murray.
Darwin, C. (1877). *The different forms of flowers on plants of the same species*. London: John Murray.
Darwin, C. (1987). *Charles Darwin's notebook, 1836–1844*. New York: Cornell University Press.
Darwin, F., & Seward, A. C. (Eds.). (1903). *More letters of Charles Darwin*. London: John Murray.
Davison, J. A. (1993). The blind alley: Its significance for evolutionary theory. *Rivista di Biologia/Biology Forum, 86*, 101–110.
Davison, J. A. (2004). Is evolution finished? *Rivista di Biologia/Biology Forum, 97*, 111–116.
de Beer, G. (1964). Mendel, Darwin and Fisher (1865–1965). *Notes and Records of the Royal Society of London, 19*, 192–226.
de Beer, G. (1965). Mendel, Darwin, and the center of science. *The Listener, 11*, 364–366.
Dobzhansky, T. (1965). Mendelism, darwinism, and evolutionism. *Proceedings of the American Philosophical Society, 109*, 205–215.
Dunn, P. M. (2003). Gregor Mendel, OSA (1822–1884), founder of scientific genetics. Fetal and Neonatal Edition 88 *Archives of Disease in Childhood*, F537–F5399.
Durand, E., Méheust, R., Soucaze, M., Goubet, P. M., Gallina, S., Poux, C., ... Castric, V. (2014). Dominance hierarchy arising from the evolution of a complex small RNA regulatory network. *Science, 346*, 1200–1205.
Edwards, A. W. (1986). Are Mendel's results really too close? *Biological Reviews of the Cambridge Philosophical Society, 61*, 295–312.
Endersby, J. (2003). Darwin on generation, pangenesis and sexual selection. In J. Hodge, & G. Radick (Eds.), *The Cambridge companion to Darwin* (pp. 69–91). Cambridge University Press.
Esposito, M. (2013). Heredity, development and evolution: The unmodern synthesis of E.S. Russell. *Theory in Bioscience, 132*, 165–180.
Fairbanks, D. J., & Abbott, S. (2016). Darwin's influence on Mendel: Evidence from a new translation of Mendel's paper. *Genetics, 204*, 401–405.
Fairbanks, D. J., & Rytting, B. (2001). Mendelian controversies: A botanical and historical review. *American Journal of Botany, 88*, 737–752.
Fisher, R. A. (1928). The possible modification of the response of the wild type to recurrent mutations. *The American Naturist, 62*, 115–126.
Fisher, R. A. (1936). Has Mendel's work been rediscovered? *Annals of Science, 1*, 115–137.
Galton, D. J. (2018). *Standing on the shoulders of Darwin and Mendel*. Boca Raton London New York: CRC Press.
Gayon, J. (2016). From Mendel to epigenetics: history of genetics. *Comptes Rendus Biologies, 339*, 225–230.
Ghiselin, M. T. (2009). *Darwin: A reader's guide*. Occasional Papers of the California Academy of Science.
Gilmartin, P. M. (2015). On the origins of observations of heterostyly in *Primula*. *New Phytologist, 208*, 39–51.
Goring, D. R. (2016). Dominance modifier: Expanding mate options. *NaturePlants, 3*, 16210.

Hackett, S., Feldheim, K., & Alvey, M. (2006). Genes and genius: The inheritance of gregor Mendel. *DNA and Cell Biology, 25*, 655–658.
Hartl, D. L., & Fairbanks, D. J. (2007). Mud sticks: On the alleged falsification of Mendel's data. *Genetics, 175*, 975–979.
Higami, Y., & Shimokawa, I. (2000). Apoptosis in the aging process. *Cell and Tissue Research, 301*, 125–132.
Hodge, M. J. S. (1985). Darwin as a lifelong generation theorist. In D. Kohn (Ed.), *The Darwinian heritage* (pp. 207–243). Princeton: Princeton University Press.
Howard, J. (2009). Why didn't Darwin discover Mendel's laws? *Journal of Biology, 8*, 15.
Huu, C. N., Kappel, C., Keller, B., Sicard, A., Takebayashi, Y., Breuninger, H., … Lenhard, M. (2016). Presence versus absence of CVP734A50 underlies the style length dimorphism in primroses. *eLife, 5*, e17956.
Jiang, X.-F., Zhu, X.-F., & Li, Q.-J. (2018). Variation in the degree of reciprocal herkogamy affects the degree of legitimate pollination in a distylous species. *AoB Plants, 10*, ply022.
Johannsen, W. (1923). Some remarks about units in heredity. *Hereditas, 4*, 133–141.
Kampourakis, K. (2013). Mendel and the path to genetics: Portraying science as a social process. *Science & Education, 22*, 293–324.
Koestler, J. (1971). *The case of the midwife toad.* New York: Random.
Koonin, E. V. (2009). The Origin at 150: is a new evolutionary synthesis in sight? *Trends in Genetics, 25*, 473–475.
Kritsky, G. (1980). Mendel and the origin of species. *Proceedings of the Indiana Academy of Science, 90*, 330–334.
Leone, C. A. (1952). Genetics: Lysenko versus Mendel. *Transactions of the Kansas Academy of Science, 55*, 369–380.
Li, J., Cocker, J. M., Wright, J., Webster, M. A., McMullan, M., Dyer, S., … Gilmartin, P. M. (2016). Genetic architecture and evolution of the S locus supergene in *Primula vulgaris*. *Nature Plants, 2*, 16188.
Liu, Y.-S. (2005). Darwin and Mendel: Who was the pioneer of genetics? *Rivista di Biologia/ Biological Forum, 98*, 305–322.
Liu, Y.-S. (2009). A new light on Yarrell's law? *Rivista di Biologia/Biology Forum, 102*, 16–20.
Liu, Y.-S., & Li, X.-J. (2016). Darwin and Mendel today: A comment on "limits of imagination: The 150th anniversary of Mendel's laws, and why Mendel failed to see the importance of his discovery for Darwin's theory of evolution". *Genome, 59*, 75–77.
Liu, Y.-S., Zhou, X.-M., Zhi, M.-X., Li, X.-J., & Wang, Q.-L. (2009). Darwin's contributions to genetics. *Journal of Applied Genetics, 50*, 177–184.
Lizana, G. B., & Sanchez, J. A. (1994). Effects of the parents' age on the level of polymorphism at the adh locus in *Drosophila melanogaster*: I. Effects on the genic and genotypic segregation of the offspring. *The Journal of Heredity, 85*, 327–331.
Lönnig, W.-E. (2017). Mendel's Paper on the Laws of Heredity (1866): Solving the Enigma of the Most Famous 'Sleeping Beauty' in Science. In *eLS (encyclopedia of Life Sciences).* Chichester: John Wiley & Sons, Ltd. DOI: 10.1002/9780470015902.a0026823.
MacBride, E. W. (1937). Mendel, morgan and genetics. *Nature, 140*, 348–350.
Mather, K. (1942). Genetics and the Russian controversy. *Nature, 149*, 427–430.
Mawer, S. (2006). Sowing the seeds. *Nature, 444*, 418.
Mayr, E. (1973). The recent historiography of genetics. *The Journal of the History of Biology, 6*, 125–154.
McClure, B. (2016). The genetic basis of heterostyly. *Nature Plants, 2*, 16184.
Meijer, O. G. (1985). Hugo de Vries no Mendelian? *Annals of Science, 42*, 189–232.
Mendel, G. (1901). Experiments in plant hybridisation. *Journal of the Horticultural Society of London, 26*, 1–32.
Michurin, I. V. (1949). *Selected works.* Moscow: Foreign Languages Publishing House.

Monaghan, F. V., & Corcos, A. F. (1984). On the origins of the Mendelian laws. *The Journal of Heredity, 75*, 67–69.
Muntzing, A. (1959). Darwin's views on variation under domestication in the light of present-day knowledge. *Proceedings of the American Philosophical Society, 103*, 190–220.
Novitski, C. E. (2004a). Revision of Fisher's analysis of Mendel's garden pea experiments. *Genetics, 166*, 1139–1140.
Novitski, E. (2004b). On Fisher's criticism of Mendel's results with the garden pea. *Genetics, 166*, 1133–1136.
Olby, R. C. (1966). *Origins of mendelism*. London: Constable.
Olby, R. (1979). Mendel no mendelian? *History of Science, 17*, 53–72.
Oldroyd, D. (1984). Gregor Mendel: Founding-father of modern genetics? *Endeavour, 8*, 29–31.
Opitz, J. M., & Bianchi, D. W. (2015). MENDEL: morphologist and Mathematician Founder of Genetics – to Begin a Celebration of the 2015 Sesquicentennial of Mendel's Presentation in 1865 of his Versuche über Pflanzenhybriden. Molecular. *Genetics & Genomic Medicine, 3*, 1–7.
Orel, V. (1996). *Gregor Mendel: The first geneticist*. Oxford: Oxford University Press.
Owens, S. J., & Miller, R. (2009). Cross- and self-fertilization of plants – Darwin's experiments and what we know now. *Botanical Journal of the Linnean Society, 161*, 357–395.
Padian, K. (2008). Darwin's enduring legacy. *Nature, 451*, 632–634.
Pagel, M. (2009). Natural selection 150 years on. *Nature, 457*, 808–811.
Piegorsch, W. W. (1990). Fisher's contributions to genetics and heredity, with special emphasis on the Gregor Mendel controversy. *Biometrics, 46*, 915–924.
Pinti, M., Cevenini, E., Nasi, M., De Biasi, S., Salvioli, S., Monti, D., ... Cossarizza, A. (2014). Circulating mitochondrial DNA increases with age and is a familiar trait: Implications for "inflamm-aging". *European Journal of Immunology, 44*, 1552–1562.
Paul, D. B. (1983). A War on Two Fronts: J.B.S. Haldane and the Response to Lysenkoism in Britain. *Journal of the History of Biology, 16*, 1–37.
Quintanilha, A. (1959). Social implications of Mendelism versus Michurinism. *Nature, 183*, 1222–1224.
Radick, G. M. (2015). History of Science. Beyond the Mendel-Fisher Controversy. *Science, 350*, 159–160.
Roberts, H. F. (1919a). A Darwinian statement of the Mendelian theory. *Nature, 103*, 463–464.
Roberts, H. F. (1919b). Darwin's contribution to the knowledge of hybridization. *The American Naturalist, 53*, 535–554.
Sandler, I. (2000). Anecdotal, historical and critical commentaries on genetics: Development : Mendel's legacy to genetics. *Genetics, 154*, 7–11.
Sermonti, G. (1985). Mendelism: A self-reproducing law. *Rivista di Biologia/Biology Forum, 78*, 7–8.
Singh, R. S. (2015). Limits of imagination: the 150th anniversary of Mendel's laws, and why Mendel failed to see the importance of his discovery for Darwin's theory of evolution. *Genome, 58*, 415–441.
Smykal, P., Varshney, R. K., Singh, V. K., Coyne, C. J., Domoney, C., Kejnovsky, E., & Warkentin, T. (2016). From Mendel's discovery on pea to today's plant genetics and breeding: commemorating the 150th anniversary of the reading of Mendel's discovery. *Theoretical and Applied Genetics, 129*, 2267–2280.
Stern, C., & Sherwood, E. R. (1966). *The origin of genetics: A Mendel source book*. San Francisco: W. H. Freeman.
Tarutani, Y., Shiba, H., Iwano, M., Kakizaki, T., Suzuki, G., Watanabe, M., ... Takavama, S. (2010). Trans-acting small RNA determines dominance relationships in Brassica self-incompatibility. *Nature, 466*, 983–987.

Theunissen, B. (1994). Closing the door on Hugo de Vries's Mendelism. *Annals of Science, 51*, 225–248.
Torres, T. T. (2016). Genetics teaching: Carry on celebrating Mendel's legacy. *Nature, 534*, 475.
Vorzimmer, P. J. (1969). Darwin's "questions about the breeding of animals". *Journal of the History of Biology, 2*, 269–281.
Wallace, A. R. (1908). The present position of Darwinism. *Contemporary Review, 94*, 129–142.
Wang, X.-M. (1998). *Explorations into genetics in broadened sense*. Wuhan: Wuhan University Press.
Weiling, F. (1986). What about R. A. Fisher's statement of the "too good" data of J. G. Mendel's Pisum paper? *The Journal of Heredity, 77*, 281–283.
Weller, S. G. (2009). The different forms of flowers – what have we learned since Darwin? *Botanical Journal of the Linnean Society, 160*, 249–261.
Yasuda, S., Wada, Y., Kakizaki, T., Tarutani, Y., Miura-Uno, E., Murase, K., ... Takavama, S. (2016). A complex dominance hierarchy is controlled by polymorphism of small RNAs and their targets. *Nature Plants, 3*, 16206.
Zhang, L., Kokkonen, G., & Roth, G. S. (1995). Identification of neuronal programmed Cell death in Situ in the striatum of normal adult rat brain and its relationship to Neuronal death during aging. *Brain Research, 677*, 177–179.

CHAPTER TWO

Darwin's Pangenesis and Graft Hybridization

Yongsheng Liu[*,§]
*Henan Collaborative Innovation Center of Modern Biological Breeding, Henan Institute of Science and Technology, Xinxiang, China
§Department of Biochemistry, University of Alberta, Edmonton, AB, Canada
E-mail: ysliu63@yahoo.ca.

Contents

Abstract

Although there were many records of graft-induced variations in ancient China, it was Darwin who coined the term "graft hybridization", the formation of hybrids between distinct species or varieties, through plant grafting, without the intervention of the sexual organs. He described many cases of the so-called "graft hybrids", in which

Advances in Genetics, Volume 102
ISSN 0065-2660
https://doi.org/10.1016/bs.adgen.2018.05.007

shoots produced from grafted plants exhibited a combination of characters of both rootstock and scion, and explained their formation by his Pangenesis. Michurin invented "mentor-grafting" and "preliminary vegetative approximation" methods, which greatly increased the production of graft hybrids, thus providing a solution to Darwin's puzzle. Over the past decides, the existence of graft hybrids has been extensively documented, and graft hybridization is considered to be a simple and efficient means of plant breeding, and would be especially significant in the improvement of fruit trees. Graft hybridization is now explained by horizontal gene transfer and DNA transformation. In addition, the long-distance transport of mRNA and small RNAs is also considered to be involved in the formation of graft hybrids.

1. DARWIN ON GRAFT HYBRIDIZATION

1.1 Darwin Coined the Term "Graft Hybridization"

The practice of grafting had been carried on by the Chinese for thousands of years (Daniel, 1929; Hartmann, Kester, Davies, & Geneve, 1997; Zhou, 1994). There were also records for graft-induced changes in China more than 1400 years ago (Liu, 2000a; Liu, Li, Li and Zhou, 2004). For example, the chapter on "Grafting pear trees" of the famous book *Qi Min Yao Shu* by Si-Xie Jia described the characters of the offspring of pear tree, which was grafted onto wild Du-li (*Pyrus betulifolia*): "There are about ten seeds in each fruit, but only two seedlings may bear fruits of good quality, while the rest are useless, like the wild Du-li." This suggests that the stock of Du-li greatly influenced the characters of the offspring of the grafted pear tree. Interestingly, this observation is still true today. The typical recorded instance for graft-induced changes in stone fruit trees was described by Shi-Zhen Li, in his book *Ben Cao Gang Mu* (*Chinese Medicinal Herbs*): "The fruits of the plum, which was grafted onto the stock of peach, have a hairy texture, if the stock of peach is much more vigorous than the scion of plum. The fruits of sour plum, which was grafted onto the stock of apricot, taste sweet and no longer sour, if the stock of apricot is much more vigorous than the scion of sour plum." (cited in Liu, 2000a) These observations were later confirmed by Michurin (1949).

Although there were many records for graft-induced changes in ancient China and other countries, Darwin (1868) was the first to investigate this phenomenon systemically and comprehensively, and coined the term "graft hybridization" and "graft hybrid", which he believed to be striking evidence in favour of his Pangenesis. In chapter 11 of his *The Variation of Animals and Plants under Domestication*, Darwin wrote:

> *I will therefore give all the facts which I have been able to collect on the formation of hybrids between distinct species or varieties, without the intervention of the sexual organs. For if, as I am now convinced, this is possible, it is a most important fact, which will sooner or later change the view of physiologists with respect to sexual reproduction*
>
> ***Darwin (1875)***

As Michie (1958) pointed out, "Sooner or later" has proved to be a long time, and the fulfillment of Darwin's conditional prophecy has not yet been complete. 150 years after Darwin put forward the concept of graft hybridization, although most biologists now recognize the existence of graft chimera and somatic hybrids produced by protoplast fusion, there has been a refusal to accept the existence of graft hybrids as Darwin defined (Liu, 2006). Moreover many researchers used to link the graft hybridization with the name of Lysenko or Lysenkoism (Lindegren, 1966; Stegemann & Bock, 2009; Wang, Jiang, & Wu, 2016), totally ignoring Darwin's pioneering work in this field. Thus it is necessary to revisit Darwin's ideas on graft hybridization.

1.2 Darwin's Description of Graft Hybrids in Plants

In the 11th chapter of *The Variation of Animals and Plants and under Domestication*, Darwin gave all the facts he collected on the formation of graft hybrids. He began with the famous case of Adam's laburnum (*Cytisus adami*), a graft hybrid produced from a bud of the purple laburnum (*Cytisus purpureus*), which was grafted onto the stock of the common laburnum (*Cytisus laburnum*). The most remarkable thing about this graft hybrid is that the same plant bears different kinds of flowers. As Darwin (1875) described:

> *To behold mingled on the same tree tufts of dingy-red, bright yellow, and purple flowers, borne on branches having widely different leaves and manner of growth, is a surprising sight. The same raceme sometimes bears two kinds of flowers; and I have seen a single flower exactly divided into halves, one side being bright yellow and the other purple; so that one half of the standard-petal was yellow and of larger size, and the other half purple and smaller. In another flower the whole corolla was bright yellow, but exactly half the calyx was purple. In another, one*

of the dingy-red wing-petals had a narrow bright yellow stripe on it; and lastly, in another flower, one of the stamens, which had become slightly foliaceous, was half yellow and half purple; so that the tendency to segregation of character or reversion affects even single parts and organs.

Another remarkable fact about this graft hybrid, as Darwin noticed, was that in its intermediate state it was quite sterile. However, when the flowers became pure yellow or pure purple they yield seed. Darwin himself raised several seedlings from this tree, and noticed that these seedlings resembled the common laburnum and were perfectly fertile. When he examined branches with yellow flowers they appeared to have fully recovered their laburnum character, whereas branches with purple flowers were not always exactly like the broom. He also found that their pollen was very imperfect and a multitude of grains were small and shriveled. He examined three plants and noticed that the ovules were likewise monstrous, the nucleus varying much in shape, and projecting irregularly beyond the proper coats.

Then Darwin introduced the famous bizzarria orange, a graft hybrid produced from a seedling which had been grafted; and after the scion had perished, the stock sprouted and formed this bizzarria. Interestingly, this tree could produce leaves, flowers, and fruit as same as both bitter orange and the citron of Florence, and likewise compound fruit, with the two kinds either blended together or segregated in various ways. This graft hybrid could be propagated by cuttings, retaining its diversified character.

Darwin gave many instances in which the stock was affected by the scion. For example, when the variegated Jessamine was grafted onto the common kind, the stock sometimes produced buds bearing variegated leaves. When some buds of a golden-variegated ash were inserted into common ashes, all died except one; but the ash stocks were affected and produced shoots with variegated leaves. When buds taken from an ash-tree with yellow leaves were inserted into common ashes, the stock was also affected and produced the *Blotched Breadalbane Ash*, a variety had been propagated and preserved its character for fifty years. It had been repeatedly proved that several species of Abutilon, on which the variegated *A. thompsonii* had been grafted, became variegated. According to Darwin (1868), many researchers considered variegation as the result of disease; and some cases might be looked at as the direct result of the inoculation of a disease or some weakness. But there was a variety of the hazel with dark-purple leaves that no one had attributed this colour to disease. When this variety was grafted on the common hazel, it sometimes coloured. It is worth noting that these cases cited by Darwin greatly inspired Ross (1872), who developed his graft theory of disease.

After citing other instances of the production of graft hybrids in grape, hyacinths and rose, Darwin concluded with a rather extensive review of the results in potatoes. For example, he cited the experiments performed by Trail, who cut about sixty blue and white potatoes into halves through the eyes or buds, and then joined them, destroying the other eyes. The result showed that "some of these united tubers produced white, and others blue tubers; some, however, produced tubers partly white and partly blue; and the tubers from about four or five were regularly mottled with the two colours." Thus from these latter cases, Darwin (1875) concluded that "a stem had been formed by the union of the bisected buds, that is, by graft-hybridization." Darwin also cited the grafting experiments by Hildebrand, who inserted buds reciprocally into both kinds, destroying the other buds. Thus he "raised two plants, and each of these produced a tuber intermediate in character between the two parent-forms." According to Darwin (1875), Fenn conducted many grafting experiments, and believed that "new, intermediate varieties can be produced by grafting the tubers." His results demonstrated that "not only were the tubers affected, some being smooth and white at one end and rough and red at the other, but the stems and leaves were modified in their manner of growth, colour and precocity. Some of these graft-hybrids after being propagated for three years still showed in their haulms their new character, different from that of the kind from which the eyes had been taken." From a host of facts, Darwin believed that the potato afforded the best evidence of the possibility of the formation of graft hybrids. His summary and deductions were so clear and logical that they are quoted in full:

> *Herr Magnus asserts with much truth that graft hybrids resemble in all respects seminal hybrids, including their great diversity of character. There is, however, a partial exception, inasmuch as the characters of the two parent forms are not often homogeneously blended together in graft hybrids. They much more commonly appear in a segregated condition—that is, in segments either at first, or subsequently through reversion. It would seem that the reproductive elements are not so completely blended by grafting as by sexual generation. But segregation, of this kind occurs by no means rarely, as will be immediately shown, in seminal hybrids. Finally it must, I think, be admitted that we learn from the foregoing cases a highly important physiological fact, namely, that the elements that go to*

the production of a new being, are not necessarily formed by the male and female organs. They are present in the cellular tissue in such a state that they can unite without the aid of the sexual organs, and thus give rise to a new bud partaking of the characters of the two parent-forms.

Darwin (1875)

1.3 Darwin's Pangenetic Explanation and His Puzzles

It should be noted that by the time Darwin put forward the concept of graft hybridization, there were many who were skeptical to accept it. Some researchers attributed the cases of graft-induced changes to simple bud-variation or what we now called mutation. Darwin (1875) did not agree with this idea. His reasons was that "when varieties are produced by simple bud-variation, they frequently present quite new characters; whereas in all the numerous cases above given, as Herr Magnus likewise insists, the graft hybrids are intermediate in character between the two forms employed." Ironically, a century later, some researchers still insisted that mutation was the main reason for the formation of so-called graft hybrid (Kraevoi, 1971; Stubbe, 1954).

Some researchers thought that Darwin believed that the fusions of vegetative nuclei occurred at the graft union and resulted in graft hybrids (Burge, Morgan, & Seelye, 2002; Marcotrigiano, 1997; Neilson-Jones, 1969; Tilney-Bassett, 1986). This is misleading. A careful reading of Darwin's *The Variation of Animals and Plants under Domestication* revealed that he explained the formation of graft hybrids only with his Pangenesis.

Darwin proposed that the gemmules "thrown off" from the cells of the stock (or scion) could be transferred into scion (or stock), and united with the cells of the scion (or stock), leading to heritable changes of the scion (or stock) and their offspring. "On any ordinary theory of reproduction the formation of graft hybrids, and the action of the male element on the tissues of the mother plant, as well as on the future progeny of female animals, are great anomalies; but they are intelligible on our hypothesis." (Darwin, 1875).

Although Darwin coined the term "graft hybridization", firmly believed the existence of graft hybrids, and provided an explanation for their formation, he failed to produce any graft hybrids in his own experiment. In his words, "the cases above given seem to me to prove that under certain unknown conditions graft hybridization can be effected." "But we do not know under what conditions this rare form of reproduction is possible." (Darwin, 1875).

Darwin's another puzzle was that graft hybrids were easy to produce in some plants such as potato, but were rare in fruit trees. As he wrote: "It is evident that graft hybrids can be made much more easily with some plants, as the potato, than with others, for instance, our common fruit trees; for the latter have been grafted by the million during many centuries, and though the graft is often slightly affected, it is very doubtful whether this may not be accounted for, merely by a more or less free supply of nutriment." He also noticed that "certain fruit-trees truly propagate their kind whilst growing on their own roots; but when grafted on other stocks, and by this process their natural state is manifestly affected, they produce seedlings which vary greatly, departing from the parental type in many characters." (Darwin, 1875) In next section, we will see how Darwin's puzzles were solved by Michurin.

2. MICHURIN'S SOLUTIONS TO DARWIN'S PUZZLES

2.1 The Invention of "Mentor-Grafting" Method

Ivan Michurin (1855—1935), like Luther Burbank, was one of the most famous plant breeders in the world (Sokolov, Savelev & Goncharov, 2015). In the course of his life, he made a great many observations and studies of plants, particularly grafted fruit trees. He created 132 commercial cultivars, eleven of them are not only still cultivated now but also are included in the National Register of Protected Plant Breeding Achievements of the Russian Federation (Goncharov & Savel'ev, 2016). He also elaborated many theoretical principles and practical methods. One of Michurin's most notable achievements was his "mentor-grafting" method, with which Darwin's "unknown conditions" of producing graft hybrids could be revealed. In his paper entitled "Use of mentors in training hybrid seedlings", Michurin (1949) wrote:

> *Let me acquaint the reader in passing with a new and very interesting method I have devised, which makes it possible for the characters and properties of young hybrid fruit tree seedlings to be partially altered at the originator's will — for seedlings to be trained, so to speak, in the direction we desire, so that their good qualities are accentuated and built up, while their tendency to develop bad, undesirable qualities is checked or even totally eliminated. ... This method is as follows. Suppose we have a well-developed six-or seven-year-old hybrid seedling which is not yet bearing, and we know that, unless measures of compulsion are taken, we shall have to wait another ten years or so for its first fruiting, as one does with hybrids among whose progenitors are varieties that perhaps reach*

fruiting age only in their twentieth year. Well, if we take three or four cuttings from a definitely high-yield variety of fruiter, and graft them onto the lower branches of this seedling's crown, not far from the truck, our seedling will, under the influence of this mentor, bear fruit within the next two years. After this the mentor scion must be cut out, otherwise the mentor variety's influence may also affect the hybrid's fruit qualities, and in the course of succeeding years this change may become permanently fixed in the new variety, which, of course, may not always be desirable. If, on the contrary, the mentor variety has properties which it would be useful to impart to the fruits of the hybrid, let us leave the scions to develop and bear fruit together with the hybrid during their first three or four bearing years. This is quite long enough for the changes introduced by the mentor to become fixed in the new variety. I have made several experiments to test the effectiveness of this method in hastening the onset of fruiting in hybrid seedlings, and the results were good each time. ... Clearly, this method can also be employed to induce many other changes in the traits and properties of hybrid varieties, for example, to increase yield and fruit size, give the fruits a more vivid colouring and better winter keeping qualities, secure a higher sugar content in the flesh, make the tree more frost-resistant, and so on.

Simply put, Michurin's "mentor-grafting" method can be understood as follows: by grafting several scions taken from old variety of fruit trees onto a young seedling's crown, the young seedling acquires characters from the old variety. Michurin (1949) emphasized repeatedly that this method could be employed effectively only on young seedlings, not on old and long-established varieties. If the stock is used as mentor plant and the scion as pupil plant, it requires: (1) the scion should be taken from the seedling which is at the early stage of development; (2) the stock must be older; (3) The leaves of the scion should be kept as less as possible, making sure that the stock is growing vigorously (Michurin, 1949). For example, in 1893, Michurin grafted a bud taken from a young Antonovka apple seedling onto the crown of a three-year-old pear wilding. 5 years later, he obtained an apple-pear graft hybrid with the pear-like shape of the fruit at the stem, which he called "Reinette Bergamotte". This variant retained the character — the pear-shaped form of the fruit near the stalk when propagated vegetatively. It was crossed with various varieties of apples, and some of the offspring bore fruits resembling a pear in shape (Naumov, 1958; Uvarov, 1966).

The "mentor-grafting" method involving annual plants consists of the following: Very young seedlings (from the cotyledonary phase to 3—5 leaved stage) are grafted onto mature stocks (two to three months old, having 20 to 30 leaves). After a graft has been accomplished it is desirable to remove the leaves as frequently as possible from the branches of the scion (the pupil plant). For the stock (mentor plant), as many as possible of the leaves and branches should be preserved. The flower buds of the experimental plants should be carefully isolated in order to avoid accidental cross-pollination, and control plants should be raised each year to test their genetic purity and stability. Usually it is the offspring seedlings that produced from selfed fruits of the scion, and not the scion itself, that produced various combinations of stock and scion characters (Glavinic, 1955; Gluscenko, 1962; Ohta & Chuong, 1975a; Pandey, 1985) (Fig. 1).

The invention of Michurin's "mentor-grafting" method provides the best solution to Darwin's puzzle. When old orchard plant varieties are propagated in the usual way by means of grafting, they hardly change at all under the influence of the wild stock. According to Michurin, the reason

Figure 1 Mentor-grafting method (A): Non-graft normal plant as control. (B): Mentor-grafting: leaves in the scion were removed to facilitate chromatin translocation from stock to scion. *Reproduced with permission from Ohta, Y. (1991). Graft-transformation, the mechanism for graft-induced genetic changes in higher plants.* Euphytica, 55, *91–99.*

was that the scion of the cultivated form was taken from an old tree which has borne fruit for a number of years, possesses such a long developed power of stability that the young, and therefore weak, two- or three-year-old stock (wilding) is naturally unable to change it. On the contrary, in the case of graft hybridization, when a scion taken from a young seedling is grafted onto the crown of an adult tree, the scion itself, as well as the seeds taken from the scion, will change its properties in the direction of the adult tree.

It should be mentioned that it was Michurin's work on graft hybridization led Hasek to make the crucial decision to choose chicken embryo as his experimental model, which turned out to be crucial for the discovery of immunological tolerance (Ivanyi, 2003). Michurin's work on graft hybridization also inspired animal breeders to investigate the effects of blood transfusion on hereditary traits, which was also known as animal vegetative hybridization (Liu, 2008).

2.2 The Invention of "Preliminary Vegetative Approximation" Method

Michurin's another legacy was that he promoted the idea of remote hybridization, of very original attempts to produce new species of plants by cross-breeding them with other species. Michurin was one of the first investigators in the history of plant breeding to use not only interspecific, but also intergeneric hybridization between remote species, and obtained many valuable plant strains. To cross least closely related species of plants, it is necessary to find ways and means to overcome the resistance to crossing. One of the important barriers to wide hybridization is rejection of pollen by a foreign style. Either the pollen is unable to germinate or the pollen tube is inhibited in the pistil before reaching the egg.

Interspecific crosses which are normally difficult or impossible to obtain can be facilitated by previous grafting between the intended sexual partners. Michurin called this method "preliminary vegetative approximation", which consists in the following: cuttings taken from one-year-old hybrid seedlings are grafted onto a branch of the crown of a mature tree of a different species or genus. The grafted cuttings continue to develop in the course of the next five or six years under the constant influence of the tree to which they have been grafted, and by virtue of this they partially change their structure. When the first flowers appear on the grafted tree, they may be pollinated by the pollen collected from the flowers of the tree on which the scions were grafted. As Michurin (1949) wrote: "I use this vegetative change as an auxiliary means to approximating two different plant species so as to obtain a sexual hybrid by crossing them in the future. This is the secret of my success in obtaining interspecific plant hybrids, such as crosses between the apricot and the plum, sweet and sour cherry, *Pyrus salicifolia* and the local cultivated pear varieties, between the various species of walnut, the various species of grapes, the various species of quince; between the pear and the mountain ash; between distant species of apple; between species of currant and other small-fruit shrubs; between pumpkins and muskmelon, musk and watermelon; between the various species of lilies; between Physalis and tomatoes, etc."

Michurin's "preliminary vegetative approximation" method was confirmed by many researchers, and some confirmatory reports and comments were published in *Nature* (Evans & Denward, 1955; Michie, 1960; Nirk, 1959). For example, Nirk (1959) reported the successful crossing of *Lycopersicon esculentum* with *L. peruvianum* by employing the reciprocal grafting of parent plants prior to hybridization. The grafting was made in the early stage of growth: the scion being at the first true leaf emergence and the stock about 7 days older. When *L. peruvianum* was used as scion and *L. esculentum* as stock, the graft was easier to achieve than the reciprocal. Hybridization was successful only when *L. esculentum* was used as maternal parent. Seed development in the hybrid fruit was superior to that observed in the offspring of ungrafted crosses. By using "preliminary vegetative approximation" method, Nirk (1959) produced 8 self-fertile interspecific hybrids, and concluded that this method provided distinct possibility for the breeding of disease-resistant tomato varieties. The prior use of grafting followed by pollination also resulted in the development of true hybrids from the cross *L. esculentum* × *L. peruvianum, L. esculentum* × *L. glandulosum* (Majid, 1964, 1966; Majid, Swaminathan, & Iyer, 1968; Rao & Ram, 1971), and other hybridization combinations (Evans, 1962; Evans and Jones, 1964; Hall, 1954).

It is well known that Chinese wolfberry is ligneous, while chili pepper is herbaceous. Although they belong to Solanaceae (eggplant) family, it is impossible to cross them. Recently, hybrid plants have been successfully obtained by using grafting and sexual hybridization. A young seedling of chili pepper was grafted onto the wolfberry tree. When the grafted seedling grew more branches and began to bloom, it was pollinated with the pollens derived from other wolfberry flowers. Interestingly, some of the pollinated flowers start to develop fruit, and the pepper fruits of the first filial generation showed traits differing from their parents in shape, length and size. After selection for two generations, a most favorite pepper fruits with big size and enriched flavor was obtained (Xu, 2015). This finding was also reported by the Section of Science & Technology, China Central Television, on November 27, 2013.

2.3 Revealing the Mystery of Heredity in Grafted Fruit Trees

Although Darwin had noted that grafted fruit trees usually produced seedlings which varied greatly, departing from the parental type in many characters, he did not know the root cause of the problem. Michurin not only made similar observation as Darwin did, but also found that it was the wild stock on which the fruit tree grated that caused the great variability. According to him, the seeds derived from the grafted fruit trees were actually the graft hybrids of the wild stock and the parental plants, being the results of both graft hybridization and sexual hybridization. As Michurin (1949) wrote:

> *When we take a grafted tree of some varieties as the maternal parent in a cross, and of the resultant seedlings only a negligible proportion bear the characters of the parent types, while the majority are simply wildings, the cause is not atavism at all, but almost exclusively the very strong and stable influence of the maternal plant's stock upon the extremely weak and unstabilized constitution of the hybrid seeds. In other words, what we get is graft hybrids of the wild stock, with only the most negligible admixture of the properties of the cultivated varieties.*

Michurin (1949) noticed that all those fruit trees that had their own roots and were ungrafted, when crossed, gave a greater number of cultivated varieties with good qualities, compared with those fruits that were grafted onto wild stock. This indicates that the plant's root system plays an important role in the formation of seeds. Based on these observations, he warned the horticulturists that those fruit trees used as parental plants in hybridization should be ungrafted trees, being standing on their own roots. In this way the numbers of new varieties with good qualities would increase four-fold.

3. EARLY EXPERIMENTS ON GRAFT HYBRIDIZATION

3.1 Daniel's Extensive Experiments on Graft Hybridization

Lucien Daniel (1856—1940) was a professor of applied botany at the Faculty of Science at Rennes University (Maton and Matron,1956). He began the study on the subject of grafting in 1890, and published his last paper in 1940. He devoted half a century to working out grafting and other horticultural operations. He described many instances in which, as a result of grafting, a shoot was produced combining characters of stock and scion, and believed that such combinations of characters in a shoot could be inherited. He also demonstrated that, following a grafting operation, new heritable variations could appear which show characters not present in either scion or stock. He possessed the specimens of the perennial varieties that he had obtained and distributed some of them to scientists who wished to study them. Daniel believed that his results not only proved the accuracy of the Lamarckian inheritance of acquired characters, but also showed that, in certain cases, grafting is a powerful agent for causing the formation of new varieties (Daniel, 1929; Neilson-Jones, 1969).

As Weiss (1930) stated, Daniel was the only botanist who had greater experimental experience in grafting plants of a great variety of families. Unlike Michurin, who mainly focused on grafted fruit trees, Daniel choose for the beginning of his studies herbaceous plants — annuals, biennials, perennials, the descendants of which he could easily control. At the International Congress of Plants Sciences in New York 1926, he reported the results obtained in his studies of grafts and of the progeny from seeds collected from the grafted plants. For example, he grafted *Allaria officinalis* on *Brassica oleracea*, shoots of turnip cabbage on a variety of headed cabbage, *Daucus carota* on the garden carrot, grafts of different races of kidney beans, *Sinapis* on *Brassica*, *Solanum dulcamara* on *Atropa belladonna*, *Solanum tuberosum* on *Solanum lycopersicum*, Artemisia and Tanacetum on Chrysanthemum, and *Helianthus annuus* on *Helianthus tuberosus*, or *H. tuberosus* on *H. annuus*. When seeds obtained from the grafted plants were planted, the progeny in certain cases retained the modifications, which were thought to have been inherited. When *H. tuberosus* was grafted on *H. annuus*, aerial tuberization occurred. These tubers were grown, and aerial tuberization reappeared in the new plants. This vegetative propagation was continued through four years, and aerial tubers were formed consistently in all of the plants (Daniel, 1929).

In 1897, Daniel developed a method of grafting called "mixed grafting", by which the influence of the stock on the scion might be increased and graft hybrids might be obtained. He suggested that a shoot of the stock just below the insertion of the scion should be allowed to develop. By doing so, the upward passage of sap was promoted and also a downward passage of elaborated sap from the shoot to the stock will be induced. Therefore a blending of characters will take place. Daniel also spent much time and efforts to study the graft hybridization in fruit trees. He opposed the viewpoint by those who regarded graft hybrid as graft chimera (Weiss, 1930). It is worth noting that, in 1959, the international colloquium on grafting was held in honor of the 100th anniversary of Daniel's birth (Dean, 1962).

3.2 Winkler's Work on Graft Chimera

Hans Winkler (1877–1945) was a German botanist, and a professor of Botany at the University of Hamburg (Noguera-Solano, Ruiz-Gutierrez and Rodriguez-Caso, 2013). Now he is most widely remembered for coining the term "genome". But he was also one of the pioneers to study graft hybrids. His early experiments were concerned with the grafting together of tomato (*Lycopersicum esculentum*) and black nightshade (*Solanum nigrum*). One of his objectives was to provide evidence that graft hybrids arose as a result of hybridization (Neilson-Jones, 1969). He made saddle grafts between tomato and nightshade. After union, the scion was removed by a horizontal cut at the junction of stock and scion. From grafts of the two very distinct plants, several types of shoots were regenerated from the callus formed at the graft union. Most of the shoots were identical to either tomato or nightshade, but some demonstrated blending characters of both stock and scion, which Winkler first thought were graft hybrids. He also obtained shoots in which the main axis was composed of tomato tissue in a sector adjacent to nightshade tissue, which he termed "graft chimera" (Bailey, 1920; Winkler, 1907).

We know that tomato has $2n = 24$ and nightshade $2n = 72$ chromosomes. Although Winkler never succeeded in obtaining a graft hybrid having $24 + 72 = 96$ chromosomes as he expected, he did produce several graft chimeras and one graft hybrid he called *Solanum darwinianum*, which had 48 chromosomes. This number was not found in either tomato or nightshade. Winkler proposed that fusion had occurred between the nuclei of the two species to give rise to a nucleus containing 96 chromosomes, which was subsequently halved to 48 by some autoregulatory process. But some researchers suggested that it might have been that of a tetraploid form of tomato. It is worth mentioning that it was Winkler who coined the term "polyploid" and "genome", and showed that polyploidy plants could be obtained by grafting (Greilhuber, Doležel, Lysák, & Bennett, 2005).

Winkler concluded that hybrids may be arranged in two groups, sexual hybrid and graft hybrid. The latter may be divided into three classes according to the theoretical possibility of their method of origin: (1) Fusion graft hybrid arising from a fusion of two somatic cells derived from distinct species, which was similar to what we now called somatic hybrids produced by protoplast fusion; (2) "Influenced" graft hybrid arising from specific influences of one graft component upon the other without cell fusion, which was similar to the notion of graft hybrids by Darwin, Michurin and Daniel; (3) Chimera, in which specifically pure cells from both graft components are combined to form a new individual (Bailey, 1920).

Although Winkler coined the term "graft chimera", he did not deny the existence of other types of graft hybrids. Now many researchers have misunderstood Winkler's original suggestion, and completely equate graft hybrid with graft chimera (Caspari and Marshak,1965). As Kaellis (1965) pointed out, this is too one-sided. It should be noted that Winkler's method of producing graft chimera is completely different from Michurin's "mentor-grafting" method. Graft chimeras are generated when an adventitious shoot apical meristem of multicellular origin arises from the graft union callus of a heterograft (Marcotrigiano, 2001), whereas Michurin's graft hybrids are produced when very young seedlings are grafted onto old plants, or scions taken from old plants are grafted onto very young seedlings, in which young seedlings are susceptible to change in the direction of the old plants. To avoid confusion, it is necessary to pay attention to their difference: chimera graft hybrid may be produced from graft union; and non-chimera graft hybrid may be produced by applying Michurin's mentor-grafting method (Hirata et al., 2001; Liu, 2006).

3.3 Burbank's Graft Hybrid of Plum and Its Confirmation

Luther Burbank (1849—1926) was the famous American horticulturist. According to Janick (2015), Burbank is introduced as follows:

> *Burbank was an outstanding horticulturist, an innovative, truly amazing plant breeder, and a self-promoter who established the concept that plant breeding could be a business. He was a controversial figure and although he cannot be considered a scientist or geneticist in the academic sense, he made outstanding contributions to plant breeding, where he was truly an artist. In his lifetime he was considered in the panoply of great American inventers who include Samuel Morse, Thomas Alva Edison, Alexander Graham Bell, Henry Ford, and Cyrus McCormick. He was known as the plant wizard and he still remains the best known American horticulturist.*

In 1893, Burbank grafted a twig derived from a purple-leaved plum onto an old Kelsey plum tree. He wanted to hybridize this grafted purple-leaved plum with some of his plums. But to his disappointment, no blossom appeared during that year. The seeds gathered from the Kelsey plum tree were sown in the next year, and one seedling with deep purple leaves was obtained among the seedlings. He believed that this variant seedling was a graft hybrid between purple-leaved plum and Kelsey plum, since the grafted twig had not bloomed, and so crossing could not have occurred in the ordinary way. Furthermore, there was also no other purple-leaved plum within thousands of miles (Burbank, 1914-1915).

To confirm the observation by Burbank, our research team grafted 57 twigs from an adult purple-leaved plum tree onto the crown of a 6-year-old Yuhuang plum tree (*Prunus salicina* Lindi) (Fig. 2). A non-grafted Yuhuang plum tree nearby served as a control. The floral buds of the purple-leaved plum were fully removed every year to avoid sexual hybridization between the two species. The seeds of the Yuhuang plum were

Figure 2 Scions of purple-leaved plum were grafted onto the crown of Yuhuang plum tree. *Reproduced with permission from Zhou, X.-M., Liu, Y.-S., & Li, X.-J. (2013). Confirmation of a purple-leaved plum graft hybrid.* Genetics and Molecular Research, 12, *710–713.*

picked in July and sown in the spring every year. The results demonstrated that no obvious changes were observed among the seedlings grown from the seeds picked up in the year that grafts were made. However, there were a few variants with purplish red leaves were found among the seedlings in the three following years (see Fig. 3 and Table 1). It should be noted that obvious changes had never been observed, either in the stock of the Yuhuang plum tree on which the scions of the purple-leaved plum had

Figure 3 Seedling of the "Yuhuang" plum on which the purple-leaved plum were grafted. Some purple-leaved variants were found among the seedlings. *Reproduced with permission from Zhou, X.-M., Liu, Y.-S., Li, X.-J. (2013). Confirmation of a purple-leaved plum graft hybrid.* Genetics and Molecular Research, 12, *710–713.*

Table 1 The Variants Among the Seedlings of Grafted and Non-grafted Yuhuang Plum

Year	Treatments	Seedlings	Variants (Purple-Leaved)	Variant Rate (%)
1999	grafted	126	0	0.0
	non-grafted	98	0	0.0
2000	grafted	19	3	15.8
	non-grafted	109	0	0.0
2001	grafted	477	11	2.3
	non-grafted	115	0	0.0
2002	grafted	163	8	4.9
	non-grafted	107	0	0.0

Reproduced with permission from Zhou, X.-M., Liu, Y.-S., & Li, X.-J. (2013). Confirmation of a purple-leaved plum graft hybrid. *Genetics and Molecular Research, 12*, 710–713.

been grafted or in the seedlings that grew from the seeds of the control of the non-grafted Yuhuang plum tree. The possibility of accidental fertilization by purple-leaved plum pollen could be rejected because the floral buds of the purple-leaved plum were fully removed before blossoming, and a control tree was used for comparison (Zhou, Liu, & Li, 2013).

4. FURTHER EVIDENCE FOR GRAFT HYBRIDIZATION

4.1 Graft Hybridization in Tomato

In the former Soviet Union, between 1930s and 1960s, many researchers performed grafting experiments, in which varieties of tomato were grafted onto other tomato varieties as well as onto species of other genera of *Solanaceae* (Morton, 1951). For example, the tomato variety of "Albino" with large and yellow fruits was grafted onto a "Mexican 353" tomato stock with small and red fruits. It was found that, after grafting, the scion of "Albino" bore fruits of various colours, including red, pink, yellow, and yellow with pink stripes. The seeds of an absolutely red fruit derived from the scion of "Albino", when planted, gave rise to F_1 plants bearing fruits of the following colours: red, bright pink, pale pink, golden yellow, pale yellow, and pale yellow with pink stripes. In the F_2 plants, the progeny of red F_1 fruits produced a range of colorations including red, bright pink, pale pink, orange, golden yellow, pale yellow, and pale yellow with pink stripes. The progeny of the bright pink F_1 fruits produced F_2 plants bearing the fruits with bright pink and pale yellow with pink stripes. When "Mexican 353" tomato was grafted onto the stock of "Albino", it was found that the "Albino" tomato produced progeny bearing red, pink and golden yellow fruits; the F_2 plants grown from red F_1 fruits produced red, pink, golden yellow and pale yellow fruits; the F_2 plants grown from golden yellow F_1 fruits produced orange and golden yellow fruits. When "Albino" tomato was grafted onto "Preserving", dwarf variety with clustered leaves and small red binocular fruits, the scion of "Albino" gave rise to F_1 plants bearing small red and sometimes binocular fruits; plants with dwarf habit and clustered leaves were also observed. Grafts involving tomato varieties other than "Albino" were made as well, and graft-induced heritable changes were also observed (Avakian & Jastreb, 1941; Hudson & Richens, 1946; Lysenko, 1954).

Gluscenko (1946) conducted extensive grafting experiments by using "Golden Queen", "Ficarazzi", "Mexican 353", "Humbert" and other tomato varieties. For example, when "Mexican 353" with red fruits was grafted onto "Golden Queen" with yellow fruits, no alternations were observed in the scion of "Mexican 353", but the stock of "Golden Queen" produced nine fruits, among which four were modified in colour from slightly reddish to a form with bright red stripes or red star-shaped patches. The control plants had only pure yellow fruits. In the first generation from the seeds of the modified fruits, some plants bearing both red and yellow fruits were obtained. It is worth mentioning that the heritable changes induced by grafting in tomato were also confirmed by others researches in the former Soviet Union, and a number of new varieties had been produced (Ananyan, 1957; Khazina, 1939). The above reports were usually regarded as "Lysenkoist" literature, and their truthfulness were suspected by Western researchers.

There were also reports on graft hybridization in tomato by researchers outside the former Soviet Union. For example, Glavinic (1955) reported that by using Michurin's "mentor-grafting" method, three characters in tomato (cut leaf rather than potato leaf, yellow fruit rather than red and short fruit rather than long) were transmitted from the variety "Kartofelisn" (stock) to the variety "GoldenTrophy" (scion). Scions were taken from young seedlings at the cotyledonary stage, and the stocks were at the flower blooming stage. Interestingly, it was the first generation seedlings, produced from selfed fruits of the scion, and not the scion itself, that produced various combinations of characters of the stock and scion. When one plant of each combination was selfed, there was a tendency to breed true for the three characters, although considerable segregation also occurred. Glavinic checked the homozygosity of these varieties for four generations before she started the grafting experiments. The buds on the scion were isolated by pergameneous bags, which were kept in place until they were removed by the fruit that had been formed in it. The parental Kartofelisni and Golden Trophy were grown as controls. From the variants induced by grafting, some new varieties were developed (Glavinic, 1961). Dean (1962) recognized that Glavinic's work satisfied the minimal requirements for experiments presuming to prove the possibility of graft hybridization, although he was suspicious of Gluscenko's claims. This study was also cited and appreciated by some Western geneticists (Kaellis, 1965; Michie, 1958; Pandey, 1976).

In Japan, Kasahara, Nakamura, and Yoneyama (1955) grafted the tomato variety "Jubilee" with orange-yellow fruit onto "New Globe" with red fruit, but only two fruits obtained from the scion. From the seeds of the two fruits, eighty six F_1 plants bore orange-yellow fruits like the scion type, except for one which bore red fruits. This variant plant also had other characters of the stock. Later, they grafted "Yellow Pear" tomato onto "Marglobe". Five fruits of the scion type were produced, and the seeds in these fruits yielded 221 plants, which bore fruits resembling either the stock or the scion (cited in Dean, 1962).

There were also negative reports in grafting experiments of tomato. For example, Sachs (1951) performed experiments on intervarietal grafting in tomato, demonstrating that there was no observable influence on leaf-shape or fruit-colour either in the year of grafting or in the following generation. Stubbe (1954) performed thousands of grafts between mutants and the original tomato varieties, but failed to obtain graft hybrids. The question arises why some investigators failed to obtain graft hybrids in their test experiments, whereas many others obtained graft hybrids in most cases. The answer is that it is not merely the grafting that counts, but also the ability skillfully to apply the mentor-grafting method. I have carefully examined the grafting methods applied by Stubbe (1954), and found that they did not follow mentor-grafting method properly. Although Sachs (1951) did apply mentor-grafting method in his tomato experiment, he only made observations in a short time. It is worth noting that successful outcomes have been rare in the short term; but most investigators working over periods of several years, were able to obtain graft hybrids (Liu, 2006).

4.2 Graft Hybridization in Eggplant

In 1955, Dr. Shinoto, the president of Japanese Society of Genetics, performed grafting experiments in eggplants as a doubter, but he was quite surprised to have obtained graft hybrids. His results were as follows: (1) The variety of Kantoao (blue fruits) was grafted onto the variety of Sinkuro (black fruits). Twenty grafts were successfully made, and 16 of them bore fruits in the branches of the scions; 9 plants bore blue fruits, and 7 plants bore black fruits, indicating that alternations occurred in the scions. (2) Sinkuro was grafted onto Kantoao; of 4 grafted plants, 2 plants demonstrated that the fruits on the stock were all blue, while in the other 2 plants, leaves and fruits of the stocks all changed into black. (3) The plants of the first generation raised from seeds obtained by selfing in the blackened fruit borne on the scion of Kantoao demonstrated segregation into the two types: 25 blue plants and 10 black plants. This report was first published in the

Japanese journal *Kagaku (Science)*, and was shortly translated into Russian and Chinese. In 1958, at the International Conference of Genetics in Montreal, Canada, Shinoto presented his interesting finding that changed the entire picture concerning the existence of graft hybrids (Lindegren, 1966). Shinoto was a thoroughly reliable geneticist with more than 30 years' experience of researches, and a man of great skill, and of the highest integrity. Although Stubbe (1954) failed to obtain graft hybrids and regarded Lysenko's grafting experiment as a fraud (Crane, 1949; Hagemann, 2002), no one doubted Shinoto's result. Later, several Japanese researchers continued to perform grafting experiments in eggplants and confirmed Shinoto's result (Hirata, 1979; 1980).

Coincidentally, in 1955, the same year Shinoto published his grafting paper, Zu & Zhao also published their grafting result on eggplant. At the beginning they always failed to obtain graft hybrids, thus realizing the importance of grafting methods and experimental materials. In their studies, they chose two eggplant varieties with evident contrasting characters: 9-leaf eggplant which was characterized by its flat, large, round fruits with purple colour and late ripening, and white eggplant, which was an early ripening variety bearing small elongated fruits with greenish colour, but it later turned white and became yellow finally. The white eggplant was grafted onto 9-leaf eggplant by using three grafting methods: cleft-grafting, bud-grafting and tender-bud-grafting. The so-called tender-bud-grafting method is described as follows: When the plant of stock grew at the height of 22–25 cm, the seeds of the scion variety began to sprout. After the cotyledons spread out, the top of the stock was cut off, and a hole on the cut surface was made. The very young seedling was cut slantwise and inserted into the hole. The results showed that the tender-bud-grafting method gave rise to a much more variations than both cleft-grafting and bud-grafting method. In the tender-bud-grafting group (see Fig. 4), the rate of the purple seedlings produced from the scion of white eggplant amounted to 19.78%, while in the cleft-grafting group it was only 6.67%. They also grafted 9-leaf eggplant onto white eggplant to study the effect of scion on the stock by using a "double-grafting" method: when the stock of white eggplant grew to 30 cm, one scion of 9-leaf eggplant was grafted on the top of the stock by cleft-grafting, another scion of 9-leaf eggplant was grafted at the side of the stock by bud-grafting. It was so arranged as to leave only one bud on the stock between the two grafted scions; the remaining buds and branches on the

Figure 4 Graft-induced variations in eggplant, in which white eggplant was grafted onto 9-leaf eggplant. First row: left: fruit of white eggplant; Right: fruit of 9-leaf eggplant; Middle: fruit in the scion of the immediate generation. From second to forth row: Variant fruits in F1 generation. *Reproduced with permission from Zu, D.-M., & Zhao, Y.-S. (1957) A study on the vegetative hybridization of some* Solanaceous *plants.* Scientia Sinica, 6, *889–903.*

stock were all removed. It was found that 68.53% purple seedlings were produced from the stock of white eggplant by using this "double-grafting" method (see Fig. 5), much higher than that by using cleft-grafting method, being only 15.13% (Zu & Zhao, 1955, 1957).

Figure 5 Graft-induced variations in eggplant, in which 9-leaf eggplant was grafted onto white eggplant. First row: left: fruit of white eggplant; Right: fruit of 9-leaf eggplant; Middle: fruit in the stock of the immediate generation. Second row: Variant fruit in F1 generation. Third-Fifth row: Variant fruits in the F2 generation. *Reproduced with permission from Zu, D.-M., & Zhao, Y.-S. (1957) A study on the vegetative hybridization of some* Solanaceous *plants.* Scientia Sinica, 6, *889–903.*

In the early 1960s, Stroun and colleagues successfully confirmed the graft-induced changes in eggplants. The chose the mentor plant from a variety characterized by a long violet fruit, brown-edged stamens, and the pupil plant from a variety with a round white fruit, yellow-edged stamen and green stem. Long violet fruit and brown-edged stems are dominant characters and there is a linkage between violet fruits and brown stamens. After grafting, all the leaves of the pupil plant were removed, the foliage of the mentor plant was kept but its flower buds were removed to avoid a sexual crossing. The flowers of the pupil plant were self-pollinated. After three successive generations of grafting, genetic modifications of the pupil plant were observed: (1) usually modifications in the pupil plant were similar to the characters of the mentor plant; (2) new characters that not observed in the mentor plant could produce in the pupil plant; (3) not all the modified pupil plants acquired the same character, some acquiring only one character, some acquiring one more characters, and some acquiring all the characters of the mentor plant; (4) sometimes the alterations appeared in mosaic form; (5) during segregation, some recessive parents produced offspring with dominant characters; (6) disjunction occurred as early as the F1 generation of the modified plant, contrary to the expectation in sexual crossing where disjunction appears in the F2 generation; (7) linked characters in the mentor plant sometimes appeared individually in the pupil plant and its offspring (Stroun & Anker, 2005, 2006; Stroun, Anker, Maurice, & Gahan, 1977; Stroun, Mathon, & Stroun, 1963).

4.3 Graft Hybridization in Pepper

During 1950s-1960s, several Soviet researchers conducted grafting experiments in pepper, and obtained new varieties (Hazina, 1956; Jagisita, 1960). In Japan, Kasahara, Nakamura, and Yoneyama (1973) confirmed the graft-induced changes in pepper. In 1967, Ohta as a Japanese "Mendelist" was permitted to examine Kasahara's original data. He realized that some anomalies could be due to mistakes but believed that other incongruities appeared potentially real and beyond the knowledge of Mendelian genetics. He and colleagues performed grafting experiments by using Kasahara's materials and methods, obtaining experimental results paralleling Kasahara's (Ohta, 1970, 1991). According to Ohta and Chuong (1975a), two varieties of red peppers used in their grafting experiments were provided by Kasahara, which had been maintained more than five generations by selfing. They chose three fruit characters for analysis: fruiting direction as being erect (up) and pendent (Up); fruiting habit or position as fasciculate (fa) or clustering and non-fasciculate (Fa) or non-clustering; and ripe pericarp colour as yellow (yc) and red (YC). Variety Tochigisantaka (T) is erect, fasciculate and red, while Kiiro (K) is pendent, non-fasciculate and yellow. The graft combination is expressed as a fraction, the numerator being a scion and the denominator being a stock. For example, T/K means a graft in which T-scion was grafted onto K-stock (Fig. 6).

Ohta and Chuong (1975a, 1975b) demonstrated that in *Capsicum annuum* Mendelian genes could be transferred from stock into scion by grafting. They grafted an immature scion, such as a small seedling 10–14 days after emergence, onto a more mature stock. Leaves of the scion, except for two or three leaves at the top, were removed twice a week during the entire growth period. When the scion was allowed to mature and set seed, plants grown from these seeds exhibited genetic characters of the stock. The overall rate of variant occurrence was 0.84%. This rate was not as high as that obtained by other researchers, who used even more young seedling at the cotyledonary stage as scion. Ohta and Chuong (1975a) summarized the characters of the graft-induced genetic changes as follows: (1) The changes are at the gene level and in both directions (from dominant to recessive and from recessive to dominant). (2) The change is usually homozygous to

Figure 6 Two parental graft combinations. K/T = Kiiro scion was grafted onto ochigisantaka stock. T/K = Tochigisantaka scion was grafted onto Kiiro stock. *Reproduced with permission from Ohta, Y., & Chuong, P. V. (1975a). Heredity changes in* Capsicum annuum *L. I. Induced by ordinary grafting. Euphytica, 24, 355–368.*

heterozygous, but sometimes it can be dominant homozygous to recessive homozygous. (3) A variant individual can carry either a single gene change or a change of several genes on different chromosomes. (4) The rate of occurrence of change varies widely from one experiment to another, and the overall rate of variant occurrence was 0.84%. (5) Variants usually appeared singly, but in one instance, all of the seeds inside a fruit gave rise to variants. (7) In the case of two complementary genes, both genes always changed together. In another set of experiments, Ohta and Chuong (1975b) reported that in stocks artificially infected with the single-stranded RNA virus for, the rate of gene transfer increased significantly from 2.0% for non-infected stocks to 16.5% for infected stocks. It appeared that the hereditary changes associated with grafting were enhanced by virus infection. It should be noted that the altered characters never appeared in the grafted generation (G_0) but only in the sexual progeny of the grafted generation. The altered traits were usually stable in the variants but sometimes demonstrated unstable chimeric mosaicism, which appeared rather often in the progeny from virus-inoculated grafts. Ohta and Chuong (1975a) found that pairs of complementary genes such as those for pericarp colour always changed together. Pandey (1976, 1985) believed that there could be no doubt that these observations are genuine, since they were made by professional Mendelian geneticists, and have been repeated with similar results by independent laboratories.

Yagishita, Hirata, and colleagues had performed grafting experiments for decades, and reported various kinds of variations with irregular genetic behaviours in the offspring after repeated grafting (Hirata & Yagishita, 1986; Hirata et al., 2003; Taller, Hirata, Yagishita, Kita, & Ogata, 1998; Taller, Yagishita, & Hirata, 1999; Yagishita & Hirata, 1986, 1987; Yagishita, 1961; Yagishita, Hirata, Mizukami, Ohashi, & Yamashita, 1990). Yagishita (1961) had established graft-induced variant lines in pepper by five successive grafting between *C. annuum* var. grossum Sendt (stock), and *C. annuum* var. fasciculatum Irish (scion). Both the stock and the scion varieties differed in many characters, and each repetition of grafting enhanced the range of variations in the variants. The variant lines had been maintained by selfing for more than 50 generations (Ajiro, Hirata, Ogata, Shiuchi, & Cui, 2004). Using these graft-induced variant lines, they studied the genetic natures of graft-induced variations (Yagishita et al., 1990). After the induction of genetic changes by repeated grafting, they selected variant lines based on variant fruit shape, and finally established stable graft hybrid lines. According to Taller et al. (1999), a line with genetic changes that was obtained from selfed seeds of the scion of an intraspecific pepper graft was used as a source of new characters in their breeding experiment. Five new variant lines and several sublines were bred by the use of this graft-induced variant strain and two other cultivars. Their breeding aim was to combine the two-lobed fruit shape that is characteristic for the graft-induced variants with thick pericarp, high sugar content, large fruit size, and significantly reduced or eliminated pungency in the new variants. The use of the three different genotypes resulted in many favorable combinations of quantitative traits, and lines were selected with the most valuable new characters. Their results showed that the new characters induced by grafting are stable, new traits, and could be served as a novel genetic source in the breeding of pepper (Fig. 7).

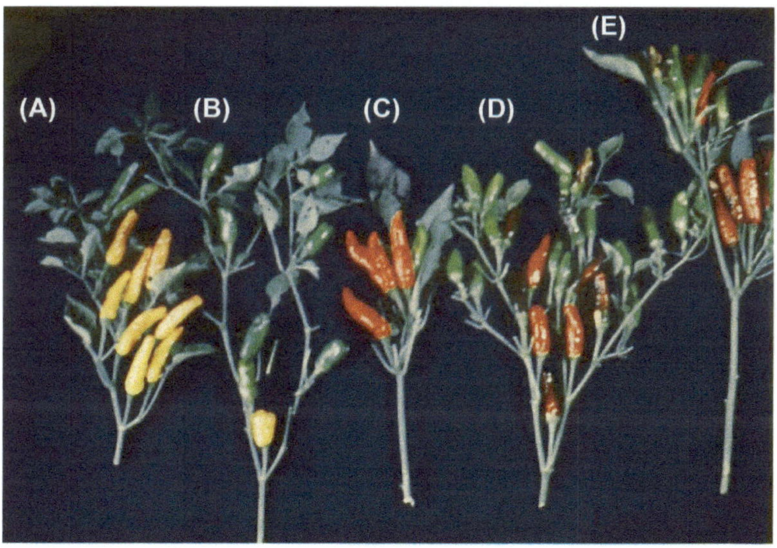

Figure 7 Change in fruit colour. The original stock (Sp) and scion (Y) cultivars as well as the graft-induced variants turn immediately from green to red during ripening. Yellow and yellow-orange fruits appeared in the G5 S3 generation of selfed graft-induced variants (A and B). Others are variations in fruit shape and fruiting habit in this generation: (C): a branch similar to 'Yatsubusa', (D): non-fasciculated, pointed and inverted-blunt fruits, (E): fasciculated, pointed and inverted-blunt fruits. *Reproduced with permission from Taller, J., Hirata, Y., Yagishita, N., Kita, M., & Ogata, S. (1998). Graft-induced changes and the inheritance of several characteristics in pepper (*Capsicum annuum *L.).* Theoretical and Applied Genetics. 97, *705–713.*

4.4 Graft Hybridization in Other Plants

During 1930s–60s, grafting experiments were also performed in potato, pea, cotton, sunflower, flax, radish, sugar beet, wheat and many other plants (Feunnov, 1938; Gluscenko, 1962; Polunia, 1950; Solodnkov, 1939). The most studies were those of Gluscenko during 1940s and 1980s (Gluscenko, 1946, 1958, 1962, 1973; Gluscenko, Bazavluk, & Medvedeva, 1948; Gluscenko, Nedesheva, & Avetisova, 1988). It was estimated that there were about 500 papers on graft hybridization published in Soviet Union during 1950–58 (Zu & Li, 1964).

Frankel (1956) grafted fertile petunias on male sterile petunia stocks, and demonstrated that some progeny of the fertile scions were male sterile, even in the progeny of second generation obtained by crossing a male sterile F1 plant with a fertile plant of the original scion. Thus he conclude that "whatever doubts may be raised by these factors of uncertainty, the fact remains that the grafting induced changes in the fertile scion that resulted in the appearance of cytoplasmic sterility in its progeny." Later he further confirmed and extended his grafting experiments (Frankel, 1962, 1971). Edwardson and Corbett (1961) also repeated this experiment and confirmed Frankel's results, showing that cytoplasmic male sterility was transmitted from the stock to the scion of petunias and into the progeny of second generation from the scion. Attempted graft transfer of cytoplasmic male sterility also succeeded in sugar beets (Cley, 1967; Corbett, 1964; Curtis, 1967) and alfalfa (Thompson & Axtell, 1978).

There was also evidence for graft-induced variations in fruit trees. An interesting phenomenon was observed in 1955 regarding the emergence of a plum shoot at the base of an old fruit-bearing branch of an apricot tree 23 cm (along the trunk) above the place where, in 1948, plum was grafted onto the apricot. Dolgushin (1957) considered it as a graft hybrid of the apricot and plum. A similar phenomenon was reported in China. In 1965, a bud derived from an adult purple-leaved plum tree was grafted onto a one-year-old apricot seedling. Next spring, the main stem of the apricot seedling was cut off above the grafting site, and the bud of the purple-leave plum began to sprout and grow. In 1980, surprisingly, an offshoot with purplish red leaves was produced from the root of the apricot and then was transplanted. The variant plant began to blossom in 1985. Interestingly, it closely resembled the scion of the purple-leaved plum but was totally different from its original plant–the stock of the apricot in all characters (Fan, 1999; Liu, 2000b).

It is interesting to introduce a mungbean variety produced by grafting. Sweet potato (*Ipomoea batatas* L.) was grown in the field, and its stems were cut off at the position 50 cm away from the growing end, being used as stock. Seeds of mungbean (*Vigna radiata* L.) were germinated in moistened soil, and young seedlings with the hypocotyls and radicles 5–7 cm in length were used as scions. A hole on the stem of sweet potato at the position 2–3 cm ascending the soil surface, and the mungbean seedling, after its radicle being cut off, was inserted into the hole of the stock. Then, the whole seedling was covered with moistened soil and shadowed immediately. The growth of the scion was maintained until the scion produced selfed seeds. The seeds were sown for several generations, and distinct genetic variations appeared in the offspring. Similar variations did not appear in the non-grafted control plants. The variations appeared to be induced by the grafting and they inherited steadily (Zhang, Meng, Xiao, Wang, & Sodmergon, 2002).

5. MOLECULAR MECHANISMS FOR GRAFT HYBRIDIZATION

5.1 Horizontal Gene Transfer and DNA Transformation

According to Darwin's Pangenesis, in grafted plants, gemmules thrown off from the cells of the rootstock (or scion) could translocate across the graft junction and unite with the gemmules thrown off from the cells of the scion (or rootstock), leading to the formation of graft hybrids. Evidently, Darwin's notion of graft hybridization is analogous to horizontal (or lateral) gene transfer (Liu, 2006). Unlike Lysenko, Michurin (1949) not only recognized the existence of genes, but also believed that genes could move between the stock and scion, resulting in the formation of graft hybrids.

The phenomenon of intercellular movement of nuclei/chromatin was first discovered in early 1900s, but it was largely thought to be an artifact. Since 1950s, several Chinese researchers rediscovered this phenomenon by using various kinds of microscopic techniques and plant materials, demonstrating that it is a naturally occurring and normal event (Lou, Wu, Chang, & Shao, 1957; Zhang, Yan, & Lou, 1990; Zheng, 1956). In 1950s and 1960s, some researchers considered intercellular movement of nuclei/chromatin as one mechanism of graft hybridization (Lou, 1963). Later, in his microhistological studies of the stock stem of pepper,

Ohta (1977) found chromatin masses, of various sizes and shapes, moving through the cell wall and intercellular space from lignifying cells towards the vascular bundles. He assumed that this chromatin must be transferred through the vascular system, across the graft junction, to the floral primordia or growing points of the scion. According to Ohta (1991), in a normal plant no genetic effects of such chromatin movements would be noticeable because all cells have the same genotype. By using "mentor-grafting" method, however, there may be considerable chromatin flow from stock to scion. Chromatin, in various sizes and shapes, observed in the lignifying cells, was considered as the donor DNA or the transforming DNA. He coined the term "graft transformation". Now it has been accepted that transformation is a probable mechanism for graft hybridization (Ajiro et al., 2004; Hirata, et al., 2003; Pandey, 1976; Taller et al., 1998, 1999).

Actually, in the early 1960s, Stroun and colleagues succeeded in inducing heritable changes in eggplants, and was the first to propose that DNA fragments could move across the graft junction and circulate throughout whole plants, and enter the somatic and reproductive cells of grafts. Later, they discovered the active release of DNA by living cells, which has been extensively studied over the past several decades (Stroun & Anker, 2006; Stroun et al., 1963, 1977). In addition, accumulating evidence has documented that cell free DNA can be secreted from mammalian cells and circulates in blood and other body fluids. There is also evidence that plants cells can release newly synthesized DNA that can freely circulate in the plants. This DNA enters cells and their nuclei where it can be integrated and expressed acting as a messenger-DNA (Gahan, 2003). In recent years, some molecular techniques have been used in analyzing graft-induced changes. For example, some RAPD makers, which were absent in the scion variety, were detected in the stock variety as well as the graft-induced variants. Some DNA fragments which were specific for stock, scion and the graft hybrids were obtained, suggesting that graft-induced variants produced from the scion have acquired specific DNA molecule from the stock (Ajiro et al., 2004; Taller et al., 1999).

Pandey (1976) proposed an alternate molecular mechanism for graft-induced heritable changes: when tissues are cut for grafting, some cells might have parts of their wall removed. This may result in fusion of stock and scion protoplasts at the graft union. If fusion involves a stock cell at interphase and a scion cell in mitosis, then chromatin of the stock cell may be pulverised. He proposed that there may also be other DNA fragmentation mechanisms that may operate in the region of the graft-interface. DNA fragments might be carried the relatively short distance up the vascular system of a seedling scion to cause transformation in the apical meristem. If certain fragments of a stock cell are taken up by a dividing cell of the scion they may be integrated into the scion chromosomes during replication. Stegemann and Bock (2009) examined grafts between tobacco plants that either expresses a transgene encoded by nuclear DNA or a transgene encoded by plastid DNA, and identified cells that express both sets of transgenic markers. They showed that genetic exchange can occur in both directions, but only involved the transfer of plastid DNA between cells, limiting to the site around the graft. Recently, Fuentes, Stegemann, Golczyk, Karcher, and Bock (2014) demonstrated that entire nuclear genomes could be transferred across the graft junction to generate a species of allopolyploid. Most recently, a new approach based on mixing and wounding of callus is being used to transfer nuclear or plastid DNA between plant cells (Sidorov, Armstrong, Ream, Ye & Saltarikos, 2018). In their grafting experimental system, however, Bock and colleagues did not observe long-distance movement of genetic materials, and only demonstrated that the genetic exchange was locally confined to the graft junction (Stegemann & Bock, 2009). This is not sufficient to explain the graft-induced changes by using Michurin's "mentor-grafting" method, as reported by many other researchers.

In human, it has been shown that after transplantation, donor-derived cell-free DNA can be detected in the blood and urine of recipient (Lo et al., 1998; Zhang et al., 1999). This approach has since been applied to plasma and urine samples for the analysis of transplantation. In plants, there is also evidence showing that cell-to-cell transfer of mitochondrial DNA across graft sites occurs rather frequently (Gurdon, Svab, Feng, Kumar, & Maliga, 2016). We know that there is striking similarity between plant grafting and human transplantation. A question may be asked: is there cell-free DNA circulating in plant sap, like those in human blood?

5.2 Long Distance Transport of mRNA and Reverse Transcription

Although the presence of endogenous mRNA in the angiosperm phloem sieve tube system has long been known, this RNA was generally considered a contaminant (Lucas, Yoo, & Kragler, 2001). Over the past several decades, grafting and transcriptomics studies have demonstrated that several thousand mRNAs move long distances from source organs to meristematic sink tissues (Ham & Lucas, 2017; Kim, Canio, Kessler, & Sinha, 2001; Morris, 2018;

Xoconostle-Cazarers et al., 1999). For example, Yang et al. (2015) found that over 3000 mRNAs were transported over long-distances in grafted grapevines. Kim et al. (2014) reported the extent to which mRNAs were exchanged during the interaction between the parasitic plant, Cuscuta, and two of its hosts, *Arabidopsis* and tomato. They identified over 9000 transcripts moving between species.

Several studies have shown that mobile mRNAs have important functions in systemic information networks for plant development. Graft transmission of tomato *Mouse ear* (*Me*) and *gibberellic acid insensitive* (*gai*) transcripts induced leaf morphological changes in wild-type scions (Chen, Janssen, Williams, & Sinha, 1997; Haywood, Yu, Huang, & Lucas, 2005; Hu et al., 2016; Kim et al., 2001). For example, Kim *et al.* reported that endogenous mRNA molecules in non-infected plants not only moved between cells, but also could execute their developmental functions in cells far removed from those in which the mRNAs were transcribed. Grafts between tomato plants with normal and mutant leaf shapes demonstrated that mRNA from the mutated gene responsible for the altered leaf shape could transfer from the mutant stock into the wild-type scion and changed leaf shape in the otherwise genotypically wild-type scion. Unfortunately, they did not investigate whether these graft-induced changes could be transmitted to future generations.

It is well known that retrotransposons are mobile genetic elements that transpose through reverse transcription of an RNA intermediate. They are ubiquitous in plants and play a major role in plant gene and genome evolution, and in many cases, retrotransposons comprise 50% of nuclear DNA content (Flavell, 1995; Kumar & Bennetzen, 1999). Retroviruses are also a potent vehicle for interspecies gene flow in plants (Peterson-Burch, Wright, & Laten, 2000). With the finding that mRNAs may transfer between cells and around the plant, and the ability of retroviruses or retrotransposons to reverse transcribe mRNA into cDNA capable of being incorporated into the genome (Adler, 2001; Spadafora, 2018; Steele, 1979; Steele, Lindley, & Blanden, 1998), it is not unreasonable to propose that an alternate mechanism exists for horizontal gene transfer from stock to scion and *vice versa* (Liu, 2004, 2006).

5.3 The Role of Small RNAs

It has been suggested that Darwin's gemmules not only include circulating cell-free DNA, but also mobiles RNAs, particularly small RNAs (Liu, 2005, 2008). In recent years, there has been accumulating evidence that small

RNAs are capable of moving between plant cells and through the vasculature (Chitwood & Timmermans, 2010; Harada, 2010; Sarkies & Miska, 2014). The mobility of small RNAs in plants have been studied by grafting experiments, in which the scion derived from a plant undergoing RNA-mediated gene silencing is grafted onto a stock plant that does not experience silencing. For instance, grafting a scion demonstrating co-suppression onto an isogenic plant in which the transgene was not silenced results in spreading of the silencing effect, indicating the existence of a long-distance silencing signal. Molnar et al. (2010) reported an elegant combination of grafting and deep sequencing, which demonstrated that 24- nucleotide long siRNAs are responsible for transmitting RNAi effects over long distances through the phloem, providing evidence that mobile small RNAs direct epigenetic modifications in the genome of the recipient cells. Moreover, a recent study have demonstrated that small RNAs from shoots of one accession transfer across a graft junction and target DNA methylation *de novo* at normally unmethylated sites in the genomes of root cells from a different accession (Lewsey et al., 2016). Most recently, Cai et al. (2018) have shown that the host plant *Arabidopsis* give off extracellular vesicles containing small RNAs, which are taken up by fungal cells, resulting in silencing of related genes during infection and altering their gene expression.

Martinez, Panda, Koheler, and Slotkin (2016) reported that identified the molecular pathway of transposable elements siRNA production in the pollen grain and showed that siRNAs produced from pollen vegetative cell transcripts could silence transposable elements reporters in the sperm cells, suggesting that transposable elements siRNAs act non-cell-autonomously, inhibiting transposable elements activity in the germ cells and potentially the next generation. Slotkin et al. (2009) analyzed transposable element activity, methylation, and mobility in developing *Arabidopsis* pollen. They demonstrated that transposons were active in the pollen vegetative nucleus and transposon-derived siRNAs seem to move from the vegetative nucleus to the sperm cells and may act to reinforce transposon silencing in the germline. In their studies on graft chimera, Li et al. (2013) showed that the change in small RNA expression induced by grafting may be an important factor for introducing graft-induced genetic changes. However, the role of small RNAs in heritable changes induced by using "mentor-grafting" has not been demonstrated yet. In animals, it has been documented that acquired characters could be inherited via small RNAs (Chen et al., 2016; Rechavi et al., 2014). A question is now pressing: is the mobile small RNAs really associated with the inheritance of characters acquired by grafting?

The subject of graft hybridization had been a central debate between Mendelian geneticists and Lysenkoists. In 1956, the primary leader of China, Zedong Mao, proposed the "Double-Hundred" slogan — "hundred flowers bloom, hundred schools contend" (Li, 1988). Thus I was taught both Mendelian genetics and Michurinian genetics when I was an undergraduate student. In 2000, I submitted a manuscript on graft hybridization to *Hereditas (Beijing)*, attempting to provide an explanation for the formation of graft hybrid obtained by Fan (1999). When I received a revision request, the editor-in-chief, Lihuang Zhu, provided me with a copy of a *Science* paper entitled "Plant paralog to viral movement protein that potentiates transport of mRNA into the phloem" (Xoconostle-Cazarers et al., 1999), which led me to link Darwin's gemmules to mobile RNAs. Now it has been shown that cells not only shed RNAs, but also DNA and entire nuclear genomes, and even extracellular vesicles, which can move from cell to cell, and transfer from one plant to another. It is to be hoped that the recent discoveries will direct the attention of geneticists to Darwin's concept of graft hybridization and his pangenetic explanation.

REFERENCES

Adler, M. J. (2001). A novel mechanism for evolution? *Science, 294*, 53—54.

Ajiro, T., Hirata, Y., Ogata, S., Shiuchi, K., & Cui, S. (2004). Molecular analysis of graft-induced variant strain in pepper (*Capsicum annuum* L). In R. E. Voorrips (Ed.), *Proceedings of the 12th EUCARPIA meeting on genetics and breeding of Capsicum and eggplant. Noordwijkerhort, Neitherkands, 17—19 May, 2004* (p. 216).

Ananyan, A. A. (1957). New tomato varieties obtained by vegetative and vegetative-field hybridization. *Agrobiology, 2*, 58—64.

Avakian, A. A., & Jastreb, M. G. (1941). Hybridization by grafting. *Vernalization, 1*, 60—77.

Bailey, L. H. (1920). *Plant breeding*. New York: The Macmillan Company.

Burbank, L. (1914—15). *His methods and discoveries and their practical application*. New York and London: Luther Burbank Press.

Burge, G. K., Morgan, E. R., & Seelye, J. F. (2002). Opportunities for synthetic plant chimeral breeding: Past and future. *Plant Cell, Tissue and Organ Culture, 70*, 13—21.

Cai, Q., Qiao, L.-L., Wang, M., He, B.-Y., Lin, F.-M., Palmquist, J., & Jin, H.-L. (2018). Plants send small RNAs in extracellular vesicles to fungal pathogen to silence virulence genes. *Science, 360*, 1126—1129.

Caspari, E. W., & Marshak, R. E. (1965). The rise and fall of Lysenko. *Science, 194*, 275—278.

Chen, J. J., Janssen, B. J., Williams, A., & Sinha, N. (1997). A gene fusion at a homeobox locus: Alterations in leaf shape and implications for morphological evolution. *The Plant Cell, 9*, 1289—1304.

Chen, Q., Yan, M.-H., Cao, Z.-H., Li, X., Zhang, Y., Shi, J., ... Zhou, Q. (2016). Sperm tsRNAs contribute to intergenerational inheritance of an acquired metabolic disorder. *Science, 351*, 397—400.

Chitwood, H., & Timmermans, C. (2010). Small RNAs are on the move. *Nature, 467*, 415—419.

Cley, G. (1967). Graft-transmission of male sterility in sugar beet. *Euphytica, 16*, 419–424.
Corbett, M. K. (1964). Inter-graft transmission of cytoplasmic male sterility. *Nature, 201*, 847–848.
Crane, M. B. (1949). Lysenko on grafting and genetics. *Discovery, 10*, 64–66.
Curtis, G. J. (1967). Graft-transmission of male sterility in sugar beet (*Beta vulgaris* L.). *Euphytica, 16*, 419–424.
Daniel, L. (1929). The inheritance of acquired characters in grafted plants. In B. M. Duggar (Ed.), *Proceedings of the international congress of plant Sciences* (pp. 1024–1044). Menasha, Wisconsin: George Banta Publishing Company.
Darwin, C. (1868). *The variation of animals and plants under domestication*. London: John Murray.
Darwin, C. (1875). *The variation of animals and plants under domestication* (2nd ed.). London: John Murray.
Dean, D. S. (1962). An evaluation of the claims of I. E. Glouchtchenko for vegetative hybridization in plants. *Journal of Heredity, 53*, 215–221.
Dolgushin, D. A. (1957). Emergence of a plum shoot on an apricot trees. *Agrobiology, 22*, 44–46.
Edwardson, J. R., & Corbett, M. K. (1961). Asexual transmission of cytoplasmic male sterility. *Proceedings of the National Academy of Sciences of the United States of America, 47*, 390–396.
Evans, A. M. (1962). Species hybridization in *Trifolium* L. Methods of overcoming species incompatibility. *Euphytica, 11*, 164–176.
Evans, A. M., & Denward, T. (1955). Grafting and hybridization experiments in the genus. *Trifolium Nature, 175*, 687–688.
Evans, A. M., & Jones, D. G. (1964). Effect of graft and sexual hybridization on the nodulation of *Trifolium ambiguum* M. B. *Annals of Botany, 28*, 221–228.
Fan, S.-Y. (1999). Phenotype variation by the action of scion *Prunus japonica* Thunb on stock *Prunus armeniaca* L. *Hereditas, 21*, 43–44.
Feunnov, F. (1938). On vegetative hybridization of potato. *Fruit and Vegetable Growing, 11*, 26–31.
Flavell, A. J. (1995). Retroelements, reverse transcriptase and evolution. *Comparative Biochemistry and Physiology B, 110*, 3–15.
Frankel, R. (1956). Graft-induced transmission to progeny of cytoplasmic male sterility in Petunia. *Science, 124*, 684–685.
Frankel, R. (1962). Further evidence on graft-induced transmission to progeny of cytoplasmic male sterility in Petunia. *Genetics, 47*, 641–646.
Frankel, R. (1971). Genetic evidence on alternative maternal and Mendelian heredity element in petunia. *Heredity, 26*, 107–119.
Fuentes, I., Stegemann, S., Golczyk, H., Karcher, D., & Bock, R. (2014). Horizontal genome transfer as an asexual path to the formation of new species. *Nature, 511*, 232–235.
Gahan, P. B. (2003). Messenger DNA in higher plants. *Cell Biochemistry and Function, 21*, 207–209.
Glavinic, R. (1955). Vegetative hybridization of tomato. In *Proceedings of the 14th international horticultural congress volume 1* (pp. 440–445). Wageningen, Netherlands: H. Veenman & Zonen.
Glavinic, R. (1961). New tomato varieties obtained by vegetative hybridization. In *Proceedings of the 15th international horticultural congress* (pp. 361–366). Netherlands: Pergamon Press.
Gluscenko, I. E. (1946). Experimental data on hybridizing tomatoes by grafting. *Agrobiology, 3*, 78–105.
Gluscenko, I. E. (1958). Some regularities of vegetative hybridization of plants. In *Proceedings of the 10th international congress of genetics (Montreal) 2* (p. 333).

Gluscenko, I. E. (1962). Vegetative hybridization and its importance for plant breeding. *Trudy Instituta Genetiki Akademiya Nauk SSR, 29*, 40–66.
Gluscenko, I. E. (1973). The nature of chimerical plants. *Plant Breeding Abstracts, 43*, 335.
Gluscenko, I. E., Bazavluk, V. J., & Medvedeva, G. B. (1948). Plant chimeras as a form of vegetative hybrids. *Trudy Instituta Genetiki Akademiya Nauk SSR, 15*, 58–97.
Gluscenko, I. E., Nedesheva, G. N., & Avetisova, L. V. (1988). Cell engineering in vivo in some vegetable crops. *Plant Breeding Abstracts, 58*, 670–671.
Goncharov, N. P., & Savel'ev, N. I. (2016). Ivan V. Michurin: On the 160th anniversary of the birth of the Russian Burbank. *Russian Journal of Genetics: Applied Research, 6*, 105–127.
Greilhuber, J., Doležel, J., Lysák, M., & Bennett, M. D. (2005). The origin, evolution and proposed stabilization of the terms 'genome size' and 'C-value' to describe nuclear DNA contents. *Annals of Botany, 95*, 255–260.
Gurdon, C., Svab, Z., Feng, Y., Kumar, D., & Maliga, P. (2016). Cell-to-cell movement of mitochondria in plants. *Proceedings of the National Academy of Sciences of the United States of America, 113*, 3395–3400.
Hagemann, R. (2002). How did East German genetics avoid Lysenkoism? *Trends in Genetics, 18*, 320–324.
Hall, O. L. (1954). Hybridization of wheat and rye after embryo transplantation. *Hereditas, 40*, 453–458.
Ham, B.-K., & Lucas, W. J. (2017). Phloem-mobile RNAs as systemic signaling agents. *The Annual Reviews of Plant Biology, 68*, 173–195.
Harada, T. (2010). Grafting and RNA transport via phloem tissue in horticultural plants. *Scientia Horticulturae, 125*, 545–550.
Hartmann, H. T., Kester, D. E., Davies, F. T., & Geneve, R. L. (1997). *Plant propagation: Principles and practices*. Upper Saddle River, N.J.: Prentice Hall.
Haywood, V., Yu, T. S., Huang, N. C., & Lucas, W. J. (2005). Phloem long-distance trafficking of gibberellic acid-insensitive RNA regulates leaf development. *The Plant Journal, 42*, 49–68.
Hazina, E. P. (1956). New varieties of sweet pepper obtained by vegetative hybridization. *Agrobiology, 6*, 47–54.
Hirata, Y. (1979). Graft-induced changes in eggplant (*S. melongena* L.) I. Changes of hypocotyl color in the grafted scions and in the progenies from the grafted scions. *Japanese Journal of Breeding, 29*, 318–323.
Hirata, Y. (1980). Graft-induced changes in eggplant (*S. melongena* L.) II. Changes of fruit color and fruit shape in the grafted scions and in the progeny of the grafted scions. *Japanese Journal of Breeding, 30*, 83–90.
Hirata, Y., Ogata, S., Kurita, S., Nozawa, G. T., Zhou, J., & Wu, S. (2003). Molecular mechanism of graft transformation in *Capsicum annuum* L. *Acta Horticulturae, 625*, 125–130.
Hirata, Y., Takeda, Y., & Motegi, T. (2001). Induction of cytoplasmic male sterility in the progenyderived from interspecific chimera between *Brassica oleracea* and *B. campestris*. *Euphytica, 117*, 143–149.
Hirata, Y., & Yagishita, N. (1986). Graft-induced changes in soybean storage proteins. I. appearance of the changes. *Euphytica, 35*, 395–401.
Hudson, P. S., & Richens, R. H. (1946). *The new genetics in the Soviet union*. Cambridge: School of Agriculture.
Hu, C., Ham, B. K., El-Shabrawi, H. M., Alexander, D., Zhang, D., & Lucas, W. J. (2016). Proteomics and metabolomics analyses reveal the cucurbit sieve tube system as a complex metabolic space. *The Plant Journal, 87*, 442–454.
Ivanyi, J. (2003). Milan Hasek and the discovery of immunological tolerance. *Nature Reviews Immunology, 3*, 591–597.

Jagisita, N. (1960). Vegetative hybridization of red peppers. *Agrobiology, 6*, 886–893.

Janick, J. (2015). Introduction to the workshop: Luther Burbank: Plant breeder, horticulturist, American hero. *HortScience, 52*, 152.

Kaellis, E. (1965). The Lysenkoists. *Science, 149*, 1443.

Kasahara, J., Nakamura, T., & Yoneyama, Y. (1955). Vegetative hybridization of tomatoes. *Journal of Faculty of Agriculture and Horticulture Iwate University, 2*, 149–160.

Kasahara, J., Nakamura, T., & Yoneyama, Y. (1973). Graft-induced variation in red peppers (Japanese). *Rec. Adv. Breed, 13*, 73–89.

Khazina, E. (1939). The changes of varietal characteristics of Solanaceae in vegetative hybridization. *Vernalization, 3*, 44–49.

Kim, M., Canio, W., Kessler, S., & Sinha, N. (2001). Developmental changes due to long-distance movement of homeobox fusion transcript in tomato. *Science, 293*, 287–289.

Kim, G., LeBlanc, M. L., Wafula, E. K., DePamphilis, C. W., Westwood, J. H., Claude, W., & Westwood, J. H. (2014). Genomic-scale exchange of mRNA between a parasitic plant and its hosts. *Science, 345*, 808–811.

Kraevoi, S. Y. (1971). The possibilities of using grafting in plant breeding. In N. P. Dubinin (Ed.), *Genetical Principles of Plant Breeding* (pp. 343–373). Moscow: Nauka.

Kumar, A., & Bennetzen, J. L. (1999). Plant retrotransposons. *Annual Review of Genetics, 33*, 479–532.

Lewsey, M. G., Hardcastle, T. J., Melnyk, C. W., Molnar, A., Valli, A., Urich, M. A., … Ecker, J. R. (2016). Mobile small RNAs regulate genome-wide DNA methylation. *Proceedings of the National Academy of Sciences of the United States of America, 113*, E803–E810.

Lindegren, C. C. (1966). *The cold war in biology*. Ann Arbor, Michigan: Planrian Press.

Liu, Y.-S. (2000a). The records of plant graft hybridization in ancient China. In Z.-J. Zhou, & G.-J. Ni (Eds.), *Agricultural history* (pp. 308–317). Nanchang: Jiangxi People's Press.

Liu, Y.-S. (2000b). The explanation for phenotype variation by the action of scion *Prunus japonica* Thunb on stock *Prunus armeniaca* L. *Hereditas, 22*, 401–402.

Liu, Y.-S. (2004). Further evidence for Darwin's Pangenesis. *Rivista di Biologia/Biological Forum, 97*, 53–66.

Liu, Y.-S. (2005). Reversion: Going back to Darwin's works. *Trends in Plant Science, 10*, 459–460.

Liu, Y.-S. (2006). Historical and modern genetics of plant graft hybridization. *Advances in Genetics, 56*, 101–129.

Liu, Y.-S. (2008). A new perspective on Darwin's Pangenesis. *Biological Reviews of the Cambridge Philosophical Society, 83*, 141–149.

Liu, Y.-S., Li, B.-Y., Li, G.-R., & Zhou, X.-M. (2004). Graft hybridization and the specificity of heredity in fruit trees. *Hereditas, 26*, 705–710.

Li, J.-X., Wang, Y., Zhang, L.-L., Liu, B., Cao, L.-W., Qi, Z.-Y., & Chen, L.-P. (2013). Heritable variation and small RNAs in the progeny of chimeras of *Brassica juncea* and *Brassica oleracea*. *Journal of Experimental Botany, 64*, 4851–4862.

Li, P.-S. (1988). Genetics in China: The Qingdao Symposium of 1956. *ISIS, 79*, 227–236.

Lo, Y. M. D., Tein, M. S., Pang, C. C., Yeung, C. K., Tong, K. L., & Hjelm, N. M. (1998). Presence of donor-specific DNA in plasma of kidney and liver transplant recipients. *Lancet, 351*, 1329–1330.

Lou, C. H. (1963). Intercellular movement of protoplasm in higher plants and its physiological significance. *Bulletin of Biology, 5*, 1–4.

Lou, C. H., Wu, S. H., Chang, W. C., & Shao, L. M. (1957). Intercellular movement of protoplasm as a means of translocation of organic material in garlic. *Scientia Sinica, 6*, 139–157.

Lucas, W. J., Yoo, B.-C., & Kragler, F. (2001). RNA as a long- distance information macromolecule in plants. *Nature Reviews Molecular Cell Biology, 2*, 849–857.

Lysenko, T. D. (1954). *Agrobiology*. Moscow: Foreign Languages Publishing House.
Majid, R. (1964). Overcoming interspecific incompatibility in *Lycopersicon*. *Current Science, 33*, 154–156.
Majid, R. (1966). Efficacy of seedling grafting for overcoming interspecific incompatibility in *Lycopersicon*. *Current Science, 35*, 420.
Majid, R., Swaminathan, M. S., & Iyer, R. D. (1968). Production and cytogenetic analysis of interspecific hybrids in *Lycopersicon*. *Indian Journal of Genetics, 28*, 275–286.
Marcotrigiano, M. (1997). Chimeras and variegation: Patterns of deceit. *HortScience, 32*, 773–784.
Marcotrigiano, M. (2001). Genetic mosaics and the analysis of leaf development. *International Journal of Plant Science, 162*, 513–525.
Martinez, G., Panda, K., Koheler, C., & Slotkin, R. K. (2016). Silencing in sperm cells is directed by RNA movement from the surrounding nurse cell. *Nature Plants, 2*, 16030.
Maton, K.-S., & Matron, C.-C. (1956). Lucien Daniel and his theory. *Agrobiology*, (6), 3–15.
Michie, D. (1958). The third stage in genetics. In S. S. Barnett (Ed.), *A century of Darwin* (pp. 56–84). London: Heinemann.
Michie, D. (1960). Interspecific hybrids of *Lycopersicum*. *Nature, 186*, 1071.
Michurin, I. V. (1949). *Selected works*. Moscow: Foreign Languages Publishing House.
Molnar, A., Melnyk, C. W., Bassett, A., Hardcastle, T. J., Dunn, R., & Baulcombe, D. C. (2010). Small silencing RNAs in plants are mobile and direct epigenetic modification in recipient cells. *Science, 328*, 872–875.
Morris, R. (2018). On the selectivity, specificity and signalling potential of the long-distance movement of messenger RNA. *Current Opinion in Plant Biology, 43*, 1–7.
Morton, A. G. (1951). *Soviet genetics*. London: Lawrence & Wishart.
Naumov, V. A. (1958). Vegetative segregation in hybrid apple seedlings. *Agrobiology, 6*, 68–74.
Neilson-Jones, W. (1969). *Plant chimeras* (2nd ed.). London: Methuen.
Nirk, H. (1959). Interspecific hybrids in *Lycorsicum*. *Nature, 184*, 1819–1820.
Noguera-Solano, R., Ruiz-Gutierrez, R., & Rodriguez-Caso, J. M. (2013). Genome: twisting stories with DNA. *Endeavour, 37*, 213–219.
Ohta, Y. (1970). A variant found in the progeny from grafting in *Cupsicum annuum*. *National Institute of Genetics, Japan: Annual Report, 20*, 34–35.
Ohta, Y. (1977). Hereditary changes induced by grafting and by DNA application and their implication in Capsicum breeding. In E. Pochard (Ed.), *'Capsicum 77', proceeding III, Eucarpia meeting on genetics and breeding of Capsicum, Avignon, Montfavet, France* (pp. 147–153).
Ohta, Y. (1991). Graft-transformation, the mechanism for graft-induced genetic changes in higher plants. *Euphytica, 55*, 91–99.
Ohta, Y., & Chuong, P. V. (1975a). Heredity changes in *Capsicum annuum* L. I. Induced by ordinary grafting. *Euphytica, 24*, 355–368.
Ohta, Y., & Chuong, P. V. (1975b). Heredity changes in *Capsicum annuum* L. II. Induced by virus-inoculated grafting. *Euphytica, 24*, 605–611.
Pandey, K. K. (1976). Genetic transformation and graft hybridization in higher plants. *Theoretical and Applied Genetics, 47*, 299–302.
Pandey, K. K. (1985). Gene movement and genetic manipulation. In M. Freeling (Ed.), *Plant genetics* (pp. 235–249). New York: Alan R. Liss, Inc.
Peterson-Burch, B. D., Wright, D. A., & Laten, H. M. (2000). Retroviruses in plants? *Trends in Genetics, 16*, 151–152.
Polunia, J. A. (1950). Vegetative hybridization of peas. *Agrobiology, 5*, 94–103.
Rao, C. H., & Ram, H. (1971). Effect of prior grafting in interspecific incompatible crosses — an assessment. *Indian Journal of Genetics and Breeding, 31*, 530–535.

Rechavi, O., Houri-Ze'evi, L., Anava, S., Goh, W. S. S., Kerk, S. Y., Hannon, G. J., ... Hobert, O. (2014). Starvation-induced transgenerational inheritance of small RNAs in *C. elegans*. *Cell, 158*, 277–287.

Ross, J. (1872). *The graft theory of disease, being an application of Mr. Darwin's hypothesis of Pangenesis to the explanation of the phenomena of the phenomena of the Zymotic diseases*. Philadelphia: Lindsay and Blakiston.

Sachs, L. (1951). 'Vegetative hybridization' in the tomato. *Nature, 167*, 282–283.

Sarkies, P., & Miska, E. A. (2014). Small RNAs break out: The molecular cell biology of mobile small RNAs. *Nature Reviews Molecular Cell Biology, 15*, 525–535.

Shinoto, Y. (1955). Graft experiments in eggplant. *Kagaku, 25*, 602–607.

Sidorov, V., Armstrong, C., Ream, T., Ye, X.-D., & Saltarikos, A. (2018). "Cell grafting"e: a new approach for transferring cytoplasmic or nuclear genome between plants. *Plant Cell Reports*. In press: doi: 10.1007/s00299-018-2292-7.

Slotkin, R., Vaughn, M., Borges, F., Tanurdzic, M., Becker, J., Feijo, J., & Martienssen, R. (2009). Epigenetic reprogramming and small RNA silencing of transposable elements in pollen. *Cell, 136*, 461–472.

Sokolov, V. V., Savelev, N. I., & Goncharov, N. P. (2015). I. V. Michurin's work on expansion of the plant horticulture assortment and improvement of food quality. *Proceedings of the Latvian Academy of Science. Section B, 69*, 190–197.

Solodnkov, E. S. (1939). Interspecific vegetative hybrids of potatoes. *Vernalization, 1*, 83–98.

Spadafora, C. (2018). The "evolutionary field" hypothesis. Non-Mendelian transgenerational inheritance mediates diversification and evolution. *Progress in Biophysics and Molecular Biology*.

Steele, E. J. (1979). *Somatic selection and adaptive Evolution: On the inheritance of acquired characters*. Toronto: Williams-Wallace.

Steele, E. J., Lindley, R. A., & Blanden, R. V. (1998). *Lamarck's Signature: How retrogenes are changing Darwin's natural selection paradigm*. Massachusetts: Perseus Books.

Stegemann, S., & Bock, R. (2009). Exchange of genetic material between cells in plant tissue grafts. *Science, 324*, 649–651.

Stroun, M., & Anker, P. (2005). Circulating DNA in higher organisms cancer detection brings back to life: An ignored phenomenon. *Cellular and Molecular Biology, 51*, 767–774.

Stroun, M., & Anker, P. (2006). Prehistory of the notion of circulating nucleic acids in plasma/serum (CNAPS): Birth of a hypothesis. *Annals of the New York Academy of Sciences, 1075*, 10–20.

Stroun, M., Anker, P., Maurice, P., & Gahan, P. B. (1977). Circulating nucleic acid in higher organisms. *International Review of Cytology, 51*, 1–48.

Stroun, M., Mathon, C., & Stroun, J. (1963). Modifications transmitted to the offspring, provoked by heterograft in *Solanum melongena*. *Archives des Sciences Genève, 16*, 1–21.

Stubbe, H. (1954). On vegetative hybridization in plants: Experiments with tomato mutants. *Kulturpflanze, 2*, 185–236.

Taller, J., Hirata, Y., Yagishita, N., Kita, M., & Ogata, S. (1998). Graft-induced changes and the inheritance of several characteristics in pepper (*Capsicum annuum* L.). *Theoretical and Applied Genetics, 97*, 705–713.

Taller, J., Yagishita, N., & Hirata, Y. (1999). Graft-induced variants as a source of novel characteristics in the breeding of pepper (*Capsicum annuum* L.). *Euphytica, 108*, 73–78.

Thompson, T. E., & Axtell, J. D. (1978). Graft-induced transmission of cytoplasmic male sterility in alfalfa. *The Journal of Heredity, 69*, 159–164.

Tilney-Bassett, R. A. E. (1986). *Plant chimeras*. London: Edeard Arnold.

Uvarov, F. Z. (1966). Pear from apple seeds. *Sadovodstvo, 2*, 31.

Wang, J., Jiang, L., & Wu, R. (2016). Plant grafting: How genetic exchange promotes vascular reconnection. *New Phytologist, 214*, 56–65.

Weiss, F. E. (1930). The problem of graft hybrids and chimaeras. *Biological Reviews of the Cambridge Philosophical Society, 5*, 231–271.
Winkler, H. (1907). Ueber Pfropfbastarde und Pflanzliche Chimaeren. *Berichte der Deutschen Botanischen Gesellschaft, 25*, 568–576.
Xoconostle-Cazarers, B., Xiang, Y., Ruiz-Medrano, R., Wang, H.-L., Monzer, J., Yoo, B. C., ... Lucas, W. J. (1999). Plant paralog to viral movement protein that potentiates transport of mRNA into the phloem. *Science, 283*, 94–98.
Xu, M. (2015). A legend of cross breeding of Chinese wolfberry with chill pepper. *Academia Journal of Agricultural Research, 3*, 1–3.
Yagishita, N. (1961). Studies on graft hybrids of *Capsicum annum* L. II.Variation in fruit shape caused by grafting of three successive generations and effects in the progeny. *The Botanical Magazine, 881*, 480–489.
Yagishita, N., & Hirata, Y. (1986). Genetic nature of bushy plant type in the variant strain induced by grafting in pepper. *Euphytica, 35*, 17–23.
Yagishita, N., & Hirata, Y. (1987). Graft-induced changes in fruit shape in pepper. I. Genetic analysis by crossing. *Euphytica, 36*, 809–814.
Yagishita, N., Hirata, Y., Mizukami, H., Ohashi, H., & Yamashita, K. (1990). Genetic nature of low capsaicin content in the variant strains induced by grafting. *Euphytica, 46*, 249–252.
Yang, Y., Mao, L., Jittayasothorn, Y., Kang, Y., Jiao, C., Fei, Z., & Zhong, G.-Y. (2015). Messenger RNA exchange between scions and rootstocks in grafted grapevines. *BMC Plant Biology, 15*, 251.
Zhang, D.-H., Meng, Z.-H., Xiao, W.-M., Wang, X.-C., & Sodmergon. (2002). Graft-induced inheritable variation in Mungbean and its application in Mungbean breeding. *Acta Botanica Sinica, 44*, 832–837.
Zhang, J., Tong, K. L., Li, P. K., Chan, A. Y., Yeung, C. K., Pang, C. C., ... Lo, Y. M. D. (1999). Presence of donor- and recipient derived DNA in cell-free urine samples of renal transplantation recipients: Urinary DNA chimerism. *Clinical Chemistry, 45*, 1741–1746.
Zhang, W. C., Yan, W. M., & Lou, C. H. (1990). Intercellular movement of protoplasm *in vivo* in developing endosperm of wheat caryopses. *Protoplasma, 153*, 193–203.
Zheng, G. C. (1956). On the process of the intercellular migration of chromatin substance and new formation of nucleus in the pollen mother cells of *Lilium Sutchuenense* Franch. *Scientia Sinica, 5*, 497–507.
Zhou, Z.-J. (1994). Origin and development of grafting in ancient China. *Studies in the History of Natural Science, 13*, 264–272.
Zhou, X.-M., Liu, Y.-S., & Li, X.-J. (2013). Confirmation of a purple-leaved plum graft hybrid. *Genetics and Molecular Research, 12*, 710–713.
Zu, D.-M., & Li, J.-G. (1964). *Vegetative hybridization in plants and animals*. Beijing: Science Press.
Zu, D.-M., & Zhao, Y.-S. (1955). A study on the vegetative hybridization of some *Solanaceous* plants. *Nongye Xuebao, 6*, 353–367.
Zu, D.-M., & Zhao, Y.-S. (1957). A study on the vegetative hybridization of some *Solanaceous* plants. *Scientia Sinica, 6*, 889–903.

CHAPTER THREE

Darwin's Pangenesis and Medical Genetics

Yongsheng Liu*,§

*Henan Collaborative Innovation Center of Modern Biological Breeding, Henan Institute of Science and Technology, Xinxiang, China

§Department of Biochemistry, University of Alberta, Edmonton, AB, Canada

E-mail: ysliu63@yahoo.ca.

Contents

Abstract

Darwin had many close connections with medical men, and made a number of observations on inherited diseases, clearly describing the inheritance of sex-linked diseases, inherited diseases at corresponding ages, infectious disease and the causes

Advances in Genetics, Volume 102
ISSN 0065-2660
https://doi.org/10.1016/bs.adgen.2018.05.008

of diseases. Darwin's Pangenesis provides an alternative explanation for cancer metastasis, and is now supported by the discovery of circulating tumour DNA. The "genometastasis hypothesis" proposed by Garcia-Olmo et al. is consistent with Pangenesis. Darwin's view of animal regeneration and his pangenetic explanation is also supported by the recent finding regarding the role of small RNAs and extracellular vesicles in regeneration. There is increasing evidence for genetic information exchange between the donor and recipient cells during transplantation, supporting Darwin's "graft hybridization" concept that gemmules released from the cells of the stock (or scion) could move into the scion (or stock) and incorporated into their cells. In addition, there is also increasing evidence for transposition of the viscera. It appears to me that there is a need to expand the concept of Darwinian medicine by incorporating Darwin's Pangenesis into it.

1. DARWIN, MEDICINE, AND INHERITED DISEASES

1.1 Darwin and Medicine

Mendelian inheritance patterns are readily recognizable as textbook examples for many single gene diseases. However, Mendelian laws seem to be inapplicable in certain cases of inherited diseases, particularly complex diseases. Barker (1922) claimed that Mendelism was "difficult if not impossible to apply to man. As someone has put it, 'the propagation of man consists of a continual crossing of polyhybrid heterozygote bastards,' not susceptible to analysis by Mendelian methods such can be applied to the study of the propagation of plants and experimental animals." Risch (2000) also pointed out that "both human genetics and epidemiology are observational as opposed to experimental sciences, and we will never be able to exert the degree of scientific control in studies of human disease that experimentalists can with model systems". Thus it is necessary to revisit Darwin's statements on inherited diseases.

As Bynum (1983) pointed out, Darwin had "medicine in his blood". Despite not being a doctor himself, Darwin had many close connections with medical men, including his father and grandfather. His grandfather Erasmus Darwin once published a book addressing the problem of inherited diseases through cases such as gout, consumption, insanity, and epilepsy. Darwin's friends and scientific interlocutors included many prominent doctors, such as the distinguished physician Sir Henry Holland, James Paget, William Roberts, Lawson Tait, John Scott Burdon-Sanderson. In a letter of 1875, Tait wrote: "the more I think over some of the problems of pathology the more I lean towards the view that their solution will be aided by regarding them from a Darwinian point of view" (cited in Méthot, 2015).

In a letter of thanks for *The Variation of Animals and Plants under Domestication*, Paget wrote: "I expect to be made even more than I am now ashamed of my ignorance (and I fear I may add that of my profession too) on the influence of inheritance on the variation and mixtures of diseases. But I hope that my deeper shame may be the beginning of deeper knowledge" (cited in Towers, 1968). In a letter to Hooker, concerning Pangenesis, Darwin wrote: "Old sir H. Holland says he has read it twice, and thinks it very tough; but believes that sooner or later 'some view akin to it' will be accepted." (Darwin, 1888) It is worth mentioning that in 1968, the centenary of the publication of Darwin's Pangenesis, in his paper on Darwin and medicine, Towers made a similar comment as Holland: "A temporarily-eclipsed hypothesis may find itself eventually rediscovered in a safe ecological niche, or even, as with those of Mendel, elevated to the heights" (Towers, 1968).

1.2 Darwin on Inherited Diseases

Darwin himself made a number of observations on inherited diseases. Appalled by the inheritance of "evil qualities", he sought comfort in the thought that "good health, vigor, and longevity are equally inherited" (Darwin, 1875). In chapter 12, Darwin wrote:

> *A long catalogue could be given of all sorts of inherited malformations and of predisposition to various diseases. With gout, fifty per cent of the cases observed in hospital practice are, according to Dr. Garrod, inherited, and a greater percentage in private practice. Everyone knows how often insanity runs in families, and some of the cases given by Mr. Sedgwick are awful,—as of a surgeon, whose brother, father, and four paternal uncles were all insane, the latter dying by suicide; of a Jew, whose father, mother, and six brothers and sisters were all mad; and in some other cases several members of the same family, during three or four successive generations, have committed suicide. Striking instances have been recorded of epilepsy, consumption, asthma, stone in the bladder, cancer, profuse bleeding from the slightest injuries, of the mother not giving milk, and of bad parturition being inherited.*
>
> ***Darwin (1875)***

Now it has been well documented that nearly all human characters, whether normal or abnormal, are inherited to a certain extent (Petronis, 2010). Many complex diseases including cancer, diabetes, and psychiatric and neurological disorders have strong heritable components. Darwin (1875) also described rinderpest as an infectious disease of cattle plague: "It has recently been ascertained that a minute portion of the mucous discharge from an animal affected with rinderpest, if placed in the blood of a healthy ox, increases so fast that in a short space of time the whole mass of blood, weighing many pounds, is infected, and every small particle of that blood contains enough poison to give, within less-than 48 h, the disease to another animal." This quotation indicates how well developed his Pangenesis was before Pasteur and Koch's germ theory of disease. Here Darwin was using germs to bolster his argument that his "gemmules" as the carriers of the hereditary information could be small enough to fit in the germ cells of higher organisms (Levin, Lipsitch, & Bonhoeffer, 1999).

Darwin (1875) maintained that "in most cases the appearance of any inherited disease is largely determined by certain critical periods in each person's life, as well as by unfavorable conditions". He consistently linked the cause of variation with changes in the environment, pointing to changed conditions of life as being responsible for variability. He held that environmental changes, acting either on the reproductive organs or the body, were necessary to generate variation (Darwin, 1868; Winther, 2000). As he wrote:

> *The greater dissimilarity of the successive children of the same family in comparison with twins, which often resemble each other in external appearance, mental disposition, and constitution, in so extraordinary a manner, apparently proves that the state of the parents at the exact period of conception, or the nature of the subsequent embryonic development, has a direct and powerful influence on the character of the offspring. ... Thus I. Geoffroy Saint-Hilaire asserts that poor women who work hard during their pregnancy, and the mothers of illegitimate children troubled in their minds and forced to conceal their state, are far more liable to give birth to monsters than women in easy circumstances.*
>
> ***Darwin (1875)***

It is now accepted that environment plays an important role in developing complex diseases. Human epidemiological studies have shown that prenatal and early postnatal environmental factors influence the adult risk of developing various chronic diseases, such as cancer, cardiovascular disease, diabetes, obesity and even behavioural disorders such as schizophrenia (Horsfall, 1962; Jirtle & Skinner, 2007; Lichtenstein et al., 2000).

1.3 Darwin on the Inheritance of Sex-Linked Diseases

Darwin (1868) was aware of a few cases regarding sex-linked inheritance. Cases are quoted on traits being transmitted from father to son but never to daughters, or from mother to daughter but never to sons. As he described:

Two brothers, their father, their paternal uncle, seven cousins, and their paternal grandfather, were all similarly affected by a skin-disease, called pityriasis versiolor; the disease, strictly limited to the males of the family (though transmitted through the females), usually appeared at puberty, and disappeared at about the age of forty or forty-five years. The second case is that of four brothers, who, when about twelve years old, suffered almost every week from severe headaches, which were relived only by a recumbent position in a dark room. Their father, paternal uncles, paternal grandfather, and grand-uncles all suffered in the same way from headaches, which ceased at the age of fifty-four or fifty-five in all those who lived so long. None of the females of the family were affected.

Darwin (1875)

Darwin (1868) also described a peculiar disorder in ten members of a family in India. This disorder appeared in some of the family's male members in each generation. It was found in the very young and involved missing and misshapen teeth, very little hair on the body and excessive dryness of the skin. Darwin noticed that the tendency to develop this disorder transmitted from the mothers to their sons, though the females were not affected. This case later was known as ectodermal dysplasia, a fascinating human genetic disease that could serve as a model for other inherited craniofacial, oral and dental diseases and disorders. Such an inheritance pattern has been identified that the mutated gene lies on the X chromosome (Slavkin, 1997).

1.4 Darwin on the Inheritance at Corresponding Ages

Darwin (1868) gave many cases under the term of "inheritance at corresponding ages", and emphasized that "large classes of diseases usually appear at certain ages, such as St. Vitus's dance in youth, consumption in early mid-life, gout later, and apoplexy still later; and these are naturally inherited at the same period." He noticed that "with certain diseases the period of inheritance occasionally or even frequently advances". He examined various cases of eye and vision illnesses, and noticed that when both parents were myopic, some of the children tended to be myopic at an earlier age or in a higher degree than their parents. He gave more instances, which is quoted as follows:

Esquirol gives several striking instances of insanity coming on at the same as that of a grandfather, father, and son, who all committed suicide near their fiftieth year. Many other cases could be given, as of a whole family who became insane at the age of forty. Other cerebral affections sometimes follow the same rule — for instance, epilepsy and apoplexy. A woman died of the latter disease when sixty-three years old; one of her daughters at forty-three, and the other at sixty-seven: the latter had twelve children, who all died from tubercular meningitis. I mention this latter case because it illustrates a frequent occurrence, namely, a change in the precise nature of an inherited disease, though still affecting the same organ.

Darwin (1875)

In a letter to *Nature*, Darwin (1881) gave another interesting instance about hair:

The hair of a gentleman of American birth (whose name I suppress) began to turn gray when he was twenty years old, and in the course of four or five years became perfectly white. He is now seventy-five years old, and retains plenty of hair on his head. His wife had dark hair, which, at the age of seventy, was only sprinkled with grey. They had four children, all daughters, now grown to womanhood. The eldest daughter began to turn grey at about twenty, and her hair at thirty was perfectly white. A second daughter began to be grey at the same age, and her hair is now almost white. The two remaining daughters have not inherited the peculiarity. Two of the maternal aunts of the father of these children 'began to turn grey at an early age, so that by middle life their hair was white.' Hence the gentleman in question spoke of the change of color of his own hair as 'a family peculiarity.'

2. DARWIN'S PANGENESIS AND CANCER METASTASIS

2.1 Darwin's Pangenesis as an Alternative Explanation of Metastasis

Cancer is a major cause of death throughout the world and metastasis causes most cancer deaths, yet our understanding of the molecular mechanism underlying this metastatic process is far from complete. In 1889, Stephen Paget proposed the "seed and soil" hypothesis and held that metastasis depended on cross-talk between "seeds" (the cancer cells) and the "soil" (specific organ microenvironment) (Fidler, 2003; Paget, 1889). Based on this hypothesis, the circulating tumour cells have long been regarded as the exclusive or predominant source of cancer metastasis. Although even small tumours can shed millions of cancer cells into the blood, only a small proportion of these cells survive on distant organs and even fewer form clinically meaningful metastases (Vanharanta & Massague, 2013). Thus circulating tumour cells are extremely rare, and it is unclear whether they are the sole source of metastasis (Plaks, Koopman, & Werb, 2013). Obviously, the "seed and soil" hypothesis cannot fully explain the origin of metastasis.

Today, our cancer biologists and most clinical oncologists seldom read Darwin in the original. The outlines of his theory of evolution are familiar to all, and the term "natural selection" has passed into everyday language. But little is known about his Pangenesis, in which it was assumed that cells could "throw off" gemmules capable of diffusion from cell to cell or circulation through the body, union with recipient cells and modification of their subsequent development, and transmission from parent to offspring. According to Darwin's Pangenesis, the primary tumour cells could release tumour gemmules into the bloodstream. If these tumour gemmules were incorporated into distant healthy cells, it would cause the spread of cancer, thus providing another source of metastasis. If these tumour gemmules were incorporated into germ cells, it would cause the transmission of cancer to next generations, indicating that cancer had a genetic basis (Liu & Li, 2012, 2016). The "tumour gemmules" may be considered as "oncogenes" or "circulating tumour DNA (ctDNA)" (Liu, 2010).

Leon, Shapiro, Sklaroff, and Yaros (1977) demonstrated that the concentration of circulating DNA is high in cancer patients compared to healthy individuals. Stroun et al. (1989) revealed that at least part of cell-free DNA in the plasma of cancer patients originated from cancer cells. The release of DNA fragments into the blood is thought to be related to the apoptosis and necrosis, as well as active secretion by living cells. Now it has been accepted that cancer cells shed tumour DNA into the blood stream, which are being investigated as a novel biomarker and a "liquid biopsy" (Schwarzenbach, Hoon, & Pantel, 2011; Wan et al., 2017). Recently, Chan et al. (2017) have sought to address whether "liquid biopsies" could be used to diagnose cancer before an individual had symptoms. They used nasopharyngeal cancer as a model, which is associated with Epstein-Barr virus infection. By screening Epstein-Barr virus DNA in the plasma from over 20,000 Chinese men, they could accurately detect early-stage nasopharyngeal cancer in this high-risk population. This indicates an exciting step in translating tumour DNA research tests into clinical use for oncology (Stewart et al. 2018). Although circulating tumour DNA has already been used as a novel biomarker, its role as a potential source of metastasis has not been fully recognized yet.

2.2 Later Metastasis Hypothesis Supporting Darwin's Pangenesis

In a 1965 paper published in *Science*, Bendich et al. first described circulating tumour DNA as a potential source of metastasis, which was at odds with the prevailing idea that circulating tumour cells are the sole source of metastasis. They proposed that DNA fragments released by tumours and circulated in the blood could be taken up by distant recipient cells, leading to a spread of the tumour (Bendich, Wilczok & Borenfreund, 1965). Klein (1981) formulated a similar idea, suggesting that the phenotypic change in certain tumours is the result of transfection of a mitotically active cell with certain tumour-derived DNA fragments. Garcia-Olmo, Garcia-Olmo, Ontanon, Martinez, and Vallejo (1999) presented their "genometastasis hypothesis", proposing that metastasis occurs through transfection of susceptible cells located in remote target organs with ctDNA containing dominant oncogenes released by the primary tumour. Obviously, these hypotheses are consistent with Darwin's pangenetic explanation (Liu, 2010; Thierry, El Messaoudi, Gahan, Anker, & Stroun, 2016).

There is increasing evidence supporting the above hypothesis. In 1944, Avery, MacLeod, and McCarty (1944) made the initial discovery of DNA-mediated transformation. This great discovery also led to the realization that DNA might play an important role in the genesis of cancer (Bendich, 1961). In the 1960s, there were several studies demonstrating that infectious DNA from the tumour-inducing virus or tumour DNA was capable of inducing tumours when injected into experimental animals or incubated with cells (Aaronson & Todaro, 1969; Bendich et al., 1965; Bendich, Borenfreund, Honda, & Steinglass, 1969). For example, Aaronson and Todaro (1969) demonstrated that DNA isolated from monkey tumour virus (SV40) could transform human fibroblast cells in tissue culture. Bendich et al. (1969) showed that heritable changes took place in some of the hamster cells when they were treated with DNA extracted from the mouse tumour cells. The cells so altered could grow in some instances as tumours in animals. Over the past several decades, numerous studies have demonstrated that cellular DNA from a variety of human and experimental animal tumours contain oncogenes capable of inducing oncogenic transformation when introduced into mammalian cells, particularly NIH 3T3 cells, by means of DNA-mediated gene transfer (see Table 1).

Most noteworthy is the study by Garcia-Olmo et al., who reported that plasma isolated from patients with *KRAS*-mutant colorectal cancers could transform NIH-3T3 cells. After the injection of these transformed NIH-3T3 cells into non-obese diabetic-severe combined immunodeficient mice, tumours containing the *KRAS*-mutation were developed (Garcia-Olmo et al., 2010). This study has been confirmed with studies using *in vivo* model system, in which cell transformation and tumourigenesis

Table 1 Evidence Supporting ctDNA as a Source of Metastasis

Donor DNA	Recipient Cells or Animals	Transfection Effects	Selected References
DNA from polyoma virus	mice	Tumour formation	Bendich et al. (1965)
Mouse tumour DNA	hamster cells	metastatic variants obtained	Bendich et al. (1969)
SV40 DNA	Human fibroblast cells	transfection succeed	Aaronson and Todaro (1969)
Mouse tumour DNA	NIH3T3 cells	Tumour formation	Shih, Padhy, and Weinberg (1981)
Human tumour DNA	NIH3T3 cells	Tumour formation	Bernstein and Weinberg (1985)
Human tumour DNA	NIH3T3 cells	Tumour formation	Thorgeirsson et al. (1985)
Kinases oncogenes	NIH3T3 cells	Tumour formation	Egan et al. (1987)
c-Ha-ras oncogene	human endothelial cells	Tumour formation	Basolo et al. (1991)
Rat tumour DNA	rat endothelial cells	metastatic variants obtained	Ke et al. (1998)
Tumour DNA	fibroblasts/endothelial cells	transfection succeed	Bergsmedh et al. (2001)
Tumour DNA	NIH3T3 cells	Tumour formation	Garcia-Olmo et al. (2010)
Human tumour DNA	NIH3T3 cells	Tumour formation	Trejo-Becerril et al. (2012)

occurred when NIH3T3 recipient cells were treated with serum of colon cancer patients and supernatant of SW480 human cancer cells (Trejo-Becerril, Perez-Cardenas & Duenas-Gonzalez, 2014; Trejo-Becerril et al., 2012). Now it has been well documented that circulating tumour DNA can indeed be taken up by healthy cells and integrated into their genomes.

2.3 The Role of Tumour-Derived Extracellular Vesicles in Metastasis

Now it is universally admitted that almost all cell types (including tumour cells) can release extracellular vesicles (EVs), which carry a cargo of proteins and nucleic acids (Becker et al., 2016; Di Liegro, Schiera, & Di Liegro, 2017). Aubertin et al. (2016) reported that one of the responses of *in vitro* and *in vivo* cancer cells to photodynamic therapy was a massive production and emission of EVs: just 1 h after the photo-activation, thousands of EVs per cell were emitted into the medium. This agrees perfectly with Darwin's assumption that many thousands gemmules must be thrown off from the cells (Darwin, 1868). Thus these EVs have been considered as one form of Darwin's so-called gemmules (Chen, Yan, & Duan, 2016). Early studies showed that platelet-derived EVs induced angiogenesis and metastasis in lung cancer (Janowska-Wieczorek et al., 2005). Now many studies have shown that tumour-secreted EVs are critical mediators of intercellular communication between tumour cells and neighbouring or distant non-tumour cells, and play an important role in cancer progression and metastasis. Although much emphasis has been placed on the potential role of RNAs packaged within EVs in regulating cancer progression and metastasis, there is increasing evidence that EVs also carry specific DNA, which is functional in recipient cells (Cai, Wu, & Jose, 2016; Kalluri & LeBleu, 2016; Kawamura, Yamamoto, Sato, & Ochiya, 2017). Thus EVs contain genetic information available for horizontal gene transfer and play a role in metastasis.

It has been proposed that EV-DNA derived from cancer cells could be transferred into potential sites of metastasis, resulting in the modulation of the cancer microenvironment and pre-metastatic niche formation. For example, Lee et al. (2014) have demonstrated that rat epithelial cell transfection by the human H-ras oncogene results in an increase in production of EVs by cancer cells. These EVs could mediate intercellular transfer of H-ras DNA to normal fibroblasts. But they further reported that EV-mediated horizontal transfection of non-transfected cells is transient, failing to produce a permanent tumorigenic conversion of these cells (Lee et al., 2016). Thus it seems that the horizontal transfer of EV-bound DNA occurs less efficiently than that of circulating cell-free DNA. The reason is probably because DNA inside extracellular vesicles shed by cells is well protected by dual lipid membranous coating (Hur et al., 2018; Kahlert et al., 2014).

2.4 Age, Apoptosis and Metastasis

It is well known that more than 60% of all cancers occur in those aged 65 and above. The most common explanation is based on Knudson's hypothesis, which states that cancer occurs more frequently as we age because time is necessary for cells to accumulate sufficient genetic mutations to push them into full-blown carcinogenesis (Knudson, 1971; Wu, Gomes, & Sinclair, 2014). Today we know that circulating nucleic acids occur ubiquitously and are bioactive in living organisms. Apoptosis is the most common form of cell death, and that apoptotic cells are an important source of circulating nucleic acids. Genomic DNA fragments originating presumably from apoptotic cells are always present in blood plasma and tissue fluids (Lichtenstein, Melkonyan, Tomei, & Umansky, 2001). It has also been extensively documented that there is an enhanced apoptosis with increasing age (Higami & Shimokawa, 2000; Jylhava et al., 2011)). In addition, there is a closer relationship of metastasis with the tumour DNA concentration in plasma than with circulating cancer cells (García-Olmo et al., 2008). It is not unreasonable to propose that the body of older person contains more apoptotic cells, which significantly increases the concentration of circulating nucleic acids, including tumour DNA. Thus the cancer metastasis occurs more frequently as we age, not only because time is necessary for cells to accumulate sufficient genetic mutations, but also because enhanced apoptosis increases the concentration of circulating tumour DNA, which enhances cancer metastasis.

3. DARWIN'S PANGENESIS AND TRANSPLANTATION

3.1 Plant Grafting and Human Transplantation

It was Darwin who first put forward the concept of "graft hybridization". According to his Pangenesis, gemmules released from the cells of the stock (or scion) could move into the scion (or stock) and incorporated into their cells, resulting in heritable changes of the scion and their progenies (Liu, 2006). It is worth noting that these cases cited by Darwin greatly inspired Ross (1872), who developed his graft theory of disease. Actually, there is a striking similarity between plant grafting and human transplantation. Now it has been shown that there is genetic information exchange (DNA, mRNA and small RNAs) between the cells of the stock and scion. Is there genetic information exchange between the donor and recipient cells during transplantation?

In 1950s, inspired by Michurin's work on plant graft hybridization, Soviet researchers conducted animal transplantation experiments to induce heritable changes and produce new animal varieties. For example, the ovaries of a black female rabbit were removed by surgical operation, and were replaced by the ovaries from a red-eyed female Albuno by means of transplantation. The transplanted black female was mated afterwards with an Albino male. Interestingly, she produced two white and eight black offspring. The appearance of black rabbits was explained by the influence of the recipient mother (Kushner, 1957). It is worth mentioning that Hasek experimented with chickens by the ingenious method of making a deliberate synchorial parabiosis of embryos via the chorioallantoic membrane. At hatching, the parabiots were separated and were found to be tolerant not only of each other's red cells, but also skin allografts. His paper was reprinted in the journal *Transplantation* in 2003, the 50th anniversary of the discovery of immunological tolerance (Hasek, 2003).

Evidence for porcine endogenous retrovirus (PERV) infection of human cells has provoked a public health debate over the proposed use of porcine xenografts to alleviate the shortage of human allografts. To examine the hypothesis that apoptotic cell death during porcine xenograft rejection is capable of fostering horizontal DNA transfer, Bisset, Boni, Lutz, and Schupbach (2007) co-cultured human cells with porcine donor cells, and found that 0.22% of the human cells contained porcine DNA, including PERV. Burghoff et al. (2008) detected a horizontal gene transfer of the EGFP (enhanced green fluorescent protein) marker via apoptotic bodies from endothelial cells to cardiomyocytes. Waterhouse et al. (2011) also reported that DNA could be horizontally transferred from hematopoietic to epithelial cell lines through phagocytosis of apoptotic bodies, resulting in both chimerism and genomic alterations. This indicates that the incessant charge of the transplant recipient with donor-DNA and its integration in host epithelium by horizontal DNA transference is indeed operative in the generation of epithelial cells with donor derived genome (Themeli, Waterhouse, Finkle, & Spyridonidis, 2011).

3.2 Cell Free DNA: An Upcoming Biomarker in Transplantation

In 1990s, it was shown that after organ transplantation, donor-derived cell-free DNA can be detected in the blood and urine of recipient (Lo et al., 1998; Zhang et al., 1999). This approach has since been applied to plasma and urine samples for the analysis of liver transplantation and kidney transplantation. The observations of increased levels of donor-derived cell-free DNA during acute rejection reveal its possible role as an early, noninvasive rejection marker (Agbor-Enoh et al., 2018; Beck et al., 2013; Gala-Lopez et al., 2018; Gielis et al., 2015; Goh, Muralidharan, Christophi, Do, & Dobrovic, 2017; Schutz et al., 2017). Various quantification techniques have been investigated for detecting donor-derived cell-free DNA, particularly the digital droplet PCR and massive parallel sequencing method (Gielis et al., 2015). Recently, targeted-sequencing has been suggested as an alternative method to assess donor-derived cell-free DNA level (Grskovic et al., 2016). More recently, Sharon et al. (2017) reported that the proportion of donor DNA in plasma can be used to monitor the success of engraftment in allogeneic bone marrow transplantation, showing that combined measurement of chimerism in the cell-free and cellular compartments can distinguish changes in donor-derived cell-free DNA due to graft loss and graft-versus-host disease. These studies greatly expand the utility of Genome transplant dynamics in transplantation.

Generally speaking, noninvasive prenatal testing can accurately determine fetal sex during pregnancy. Recently, however, Neofytou et al. (2018) have reported an exceptional case where noninvasive prenatal testing contradicts the ultrasound-based sex determination. When the pregnant woman was recipient of a liver transplant from a male donor, graft-derived cell-free DNA released into the maternal circulation shadowed the sex determination, indicating that noninvasive prenatal testing is not reliable when the pregnant mother undergoes the organ transplantation.

3.3 The Role of Extracellular Vesicles in Transplantation

Donor cells not only release cell free DNA, but also extracellular vesicles after transplantation of allogeneic organs and tissues. Thus recipient antigen presenting cells take up these extracellular vesicles and present donor major histocompatibility antigens on their surface, activating some alloreactive T cells via semi-direct pathway of allorecognition (Gonzalez-Nolasco, Wang, Prunevieille, & Benichou, 2018). It has been shown that extracellular vesicles carrying noninherited maternal antigens are associated with maternal microchimerism and tolerance in the progeny (Bracamonte-Baran et al., 2017). The potential use of extracellular vesicles as modulators of alloimmunity and biomarkers of rejection in allotransplantation were also described by several researchers (Gonzalez-Nolasco et al., 2018). In the allogeneic stem cell transplant setting, extracellular vesicles have also been reported as potential biomarkers of acute graft versus host disease (Lia et al., 2018).

4. DARWIN'S PANGENESIS AND REGENERATIVE MEDICINE

4.1 Darwin on Regeneration

The ability of animals to regenerate missing parts is a remarkable phenomenon of biology. Since the first scientific publications on regeneration in the early 1700s, scientists have studied this phenomenon in a wide varies of species. Regeneration in animals has also astounded Darwin and many other great thinkers (Alvarado, 2009).

Darwin (1875) believed that "a multitude of the lower animals and some vertebrates possess this wonderful power." He gave some instances: "Spallanzani cut off the legs and tail of the same salamander six times successively," and "Bonnet did so eight times; and on each occasion the limbs were reproduced on the exact line of amputation." He also cited an allied animal, the axolotl, when "had a limb bitten off, which was reproduced in an abnormal condition, but when this was amputated it was replaced by a perfect limb." From many facts Darwin reached three conclusions: (1) lower animals could regenerate lost parts far more easily than higher animals; (2) the power of regeneration was much greater during the youth of an animal or during the earlier stages of its development than during maturity; (3) the power of regeneration was diffused throughout the whole body.

Although natural selection might explain the emergence of regenerative power because the regeneration of injured parts could be useful to the survival of the organism, Darwin felt that both the formation of structures during the course of normal development and their replacement following injury to the adult were due to the workings of inheritance. According to his Pangenesis, the regeneration of lost parts is possible because the gemmules for the lost parts were produced prior to the loss and are present in the rest of the body. For example, if the leg of a salamander has been removed, the leg gemmules, which are present in the body, can migrate to the cut surface, unite with the cells of the cut surface and develop into a new limb, identical to the old one. It has been suggested that Darwin's gemmules could include mobile RNAs, particularly small RNAs (Liu, 2008).

4.2 The Role of Small RNAs in Regeneration

The early experiments on the metabolism of nucleic acids began in late 1940s (Bodian, 1947). Newman and Grossman (1951) reported that the rate of liver regeneration in rats was significantly accelerated by adding RNA or DNA to the diet. The addition to the diet of tissues rich in nucleic acids, such as spleen and kidney, was also effective in speeding liver regeneration. Williamson and Guschlbauer (1961, 1963) found that almost three times as much RNA as DNA was deposited in the regenerating wound tissue. It appeared that RNA was formed most rapidly during the very early phases of regeneration, whereas DNA was formed most rapidly at about 8 days.

In early 1980s, changes in small RNAs during nerve regeneration were observed. There was an increase in ribosomal RNA synthesis, increased synthesis of a species of small nuclear RNA, which might be involved in the formation or processing of mRNA, and the synthesis of a more heterogeneous pattern of mRNA, representing a metabolic recovery of the neuron from an acute injury phase of the reaction with a better defined effort at regeneration (Austin, Moses, & Langford, 1983). In recent years, small RNAs have been shown to have a large impact on regeneration (Jayawardena et al., 2012; Thatcher & Patton, 2010). It has been proposed that targeting specific microRNAs is a rational strategy for cardiac regenerative therapy (Hodgkinson, Kang, Dal-Pra, Mirotsou, & Dzau, 2015). For example, Eulalio et al. (2012) reported that certain small RNAs can stimulate the regeneration of heart muscle cells. They identified 40 small RNAs that significantly increased DNA synthesis and cytokinesis, and selected two of these (has-miR-590 and has-miR-199a) for further analysis. When delivered directly to the rat heart, the two small RNAs increased cardiomyocyte proliferation. In addition, they boosted the normally ineffective process of heart repair after infarction in mice. Recently, Lesizza et al. (2017) from the same laboratory have made another important observation that delivery of these synthetic small RNAs to cardiac tissue in mice can efficiently induce cardiomyocytes proliferation and regeneration. It will positively affect cardiac regenerative medicine by demonstrating the use of synthetic miRNA as potential novel therapeutic targets for cardiac regeneration after injury (Zangi & Hajjar, 2017). Recently, Kong, Nuo, Zhu, Han, and Wang (2013) discovered a group of non-platelet RNA-containing particles (NPRCPs), which could aggregate, fuse and become non-nucleated cells when co-cultured with nucleated cells *in vitro*. These NPRCP-derivd cells are able to regenerate mouse kidney tissue damaged by ischaemia.

It has been shown that several microRNAs also play an important roles in neural development and axon regeneration (Mahar & Cavalli, 2018). For example, in CAST/Ei mice, which have an increased neuronal regenerative capacity, Lisi et al. (2017) have recently described the importance of miR-7048-3p, a microRNA that is downregulated following peripheral injury. This downregulation results in an increase in expression of its target achaete-scute homologue 1 (ASCL1), a transcription factor capable of promoting growth. Most recently, Watanabe, Yamaji, and Ohtsuki (2018) have attempted to identify novel microRNAs in relation to neuronal differentiation via microarray analysis in the human neuronal differentiation model neuroblastoma SH-SY5Y cells. They have identified 15 upregulated and 8 downregulated miRNAs, and demonstrated that one of the up-regulated miRNAs, miR-664a-5p, could promote neuronal differentiation of SH-SY5Y cells.

4.3 The Role of Extracellular Vesicles in Regeneration

It has been shown that cells also secret extracellular vesicles containing microRNAs in a stable form as a cargo, which can be delivered to other cells (Valadi et al., 2007). These extracellular vesicles could protect microRNAs from endogenous RNases, and allow them to reach their target cells (Mittelbrunn & Sanchez-Madrid, 2012). It has been proposed that either extracellular vesicles engineered with artificial microRNAs or extracellular vesicles carrying microRNAs from an organ tissue may provide a therapeutic strategy to promote tissue and organ regeneration (Hayashi & Hoffman, 2017; Hutcheson & Aikawa, 2018). It has also been indicated that, upon injury to the central nervous system, extracellular vesicles may propagate inflammation across the blood—brain barrier, mediating neuroprotection and modulating regenerative processes (Holm, Kaiser, & Schwab, 2018). Mesenchymal stem cells (MSCs) are multipotent stem cells that have attracted significant attention in regenerative medicine. Recently, Keshtkar, Azarpira, and Ghahremani (2018) have discussed MSC-derived extracellular vesicles and their therapeutic potential in various diseases.

5. DARWIN'S PANGENESIS AND ORGAN TRANSPOSITION

5.1 Darwin on Transposition

Organ transposition, here Darwin meant "organs which are abnormally transposed or multiplied". He gave many examples of this phenomenon. He cited Elliott Coues, who recorded a curious case of a monstrous chicken with a perfect additional right leg articulated to the left side of the pelvis. He noticed that "Goldfish often had supernumerary fins placed on various parts of their bodies. When the tail of a lizard is broken off, a double tail is sometimes reproduced; and when the foot of the salamander was divided longitudinally by Bonnet, additional digits were occasionally formed." He also noticed analogous cases occurred with plants. For example, "supernumerary petals, stamens, and pistils, are often produced. I have seen a leaflet low down in the compound leaf of *Vicia sativa* replaced by a tendril; and a tendril possesses many peculiar properties, such as spontaneous movement and irritability. ... Again, the edge of a petal may occasionally be found including one of the highestproducts of the plant, namely, pollen; for instance, I have seen the pollen-mass of an Ophrys, which is a very complex structure, developed in the edge of an upper petal. The segments of the calyx of the common pea have been observed partially converted into carpels, including ovules, and with their tips converted into stigmas. ... Numerous analogous facts could be given."

According to Darwin's Pangenesis, "the gemmules of the transposed organs become developed in the wrong place, from uniting with wrong cells or aggregates of cells during their nascent state; and this would follow from a slight modification in their elective affinities." (Darwin, 1875)

5.2 Increasing Evidence for Transposition of the Viscera

Aristotle was the first to describe the transposition of the viscera in animals, and Fabricius described the first case of transposition of viscera in human being in 1600. There have been over 1000 cases reported in the medical literature (Leininger & Gibson, 1950). The reported frequency differs. In necropsy material it was found to be 1 in 6000 and in a mass chest survey 1 in 6500 (Arge, 1960; Blegen, 1949). Torgersen (1946) reported that the frequency was 1 in 8000−10000. It has also been suggested that the general incidence of transposition of the viscera is approximately 0.014% (Leininger & Gibson, 1950).

Cases of transposition of viscera are not only anatomical curiosities, but also are of clinical interest, serving very useful purposes of putting the physician on guard for these anomalies. Although the transposition of viscera does not contribute to ill health, its early recognition is of great importance in the treatment of many diseases, both medical and surgical. Serious mistakes may be made if the condition is not timely diagnosed (Brassett & Ellis, 1991; Parsonnet, 1929). There are many cases where ignorance of the presence of the condition has led to wrong diagnosis and wrong treatment and even disaster (Anderson, 1926).

Transposition of the viscera may be complete or partial. For example, Gannett (1889) gave an admirable description of transposition of all the organs, and made a most thorough and exhaustive post-mortem examination. The individual he described was an Italian about twenty-five years of age. His heart was in the right half of the chest cavity, and its shape was exactly the reverse of the normal: the right ventricle lying to the left, the left ventricle to the right. The left lung had three lobes, and the right lung was two-lobed, and the lower anterior tip of the upper lobe ended in a distinct lingula. The liver was situated in the left half of the abdominal cavity, exactly opposite to where it should lie, and its shape was the reverse of the normal. The spleen was upon the right side and just under the diaphragm. The cardiac end of the stomach was upon the right, the pyloric upon the left. The duodenum was upon the left. The complete transposition of viscera was also reported by many later researchers (Allan, 1910; Kapur, 1924; Liddon, 1904; Parsonnet, 1929). Torgersen (1946) collected 129 cases, 123 of them were complete.

In addition to complete transposition of the viscera, there are also partial transpositions. For example, there were cases in which the heart was the only organ transposed, and other organs were normal (Aitken, 1917; Risdon, 1906). According to Cockayne (1938), there are three kinds of partial transposition: (1) Partial transposition involving both thoracic and abdominal organs. (2) Partial or complete transposition of the abdominal, without transposition of the thoracic viscera. (3) Dextrocardia, with the chambers of the heart forming a mirror image of the normal. The relationship of these three forms to complete transposition and to one another is unknown.

5.3 The Genetic Study of Transposition

Although there are many cases of transposition of the viscera, its mechanism is still unclear. In his book *Problems of Genetics*, referring to monozygotic twins, Bateson (1913, p. 44) wrote: "If anyone could show how it is that neither of a pair of twins has transposition of the viscera the whole mystery of division would, I expect, be greatly illuminated." Many hypotheses have been formulated to explain the causation of transposition of the viscera. For example, Foerster (1861) found that in the majority of double monsters, one had complete or partial transposition of the viscera, thus proposing that in cases of transposition in separate individuals there had been monozygotic twinning with the subsequent death and absorption of the twin with normal viscera. If this were true there should be many pairs of twins, one with normal and the other with transposed viscera, but this is extremely rare (Cockayne, 1938).

Cockayne (1938) held that complete transposition of the viscera was inherited as a recessive character and was determined by a single autosomal gene. The reason was its familial incidence and general distribution within a family, its occurrence in both members of a pair of monozygotic twins, and the high percentage of first cousin marriages causing it. Tihen, Charles, and Sippel (1948) reported transposition of the viscera as being caused by a recessive gene in mice. Campbell (1963) admitted that recessive inheritance played an important part, but did not agree entirely with the single gene hypothesis.

In their recent paper entitled "From Mendel to Medical genetics", Kristoffersson and Macek (2017) stated that "only a few years after the rediscovery of Mendel's laws of inheritance in 1900, the first human genetic disorders and variants were described." In their paper, Darwin's name and his Pangenesis were not mentioned. In fact, Darwin described many cases of inherited diseases, and other phenomena pertaining to medical genetics. Mendel himself narrowly focused on plant hybridization, and never studied the inheritance of diseases. In medical genetics, so-called "complex traits" are defined as non-Mendelian traits, which are usually attributed to complex interactions between more than one genetic and/or environmental factor (Sherman, 1997). According to Hackam (2000), there are at least ten examples of non-Mendelian inheritance of human diseases, such as retinoblastoma, Alport syndrome, Machado-Joseph disease, and myotonic dystrophy. 20 years ago, Ostrer (1998) published a book *Non-Mendelian Genetics in Humans*, which addressed the basic mechanisms for the inheritance of genetic disorders in humans, and explored the evidence for a number of non-Mendelian genetic processes, including gonadal and somatic mosaicism, sex-linked inheritance, mitochondrial transmission, genomic imprinting, accelerated rates of mutation, and viral infection. It has been proposed that non-Mendelian inheritance may be more common than is supposed (Beatty, 1997).

In 1991, Williams and Nesse (1991) published an important paper entitled "The Dawn of Darwinian Medicine", which called the dearth of evolutionary biology in medical schools "unfortunate", asking physicians to be "as attuned to Darwin as they have been to Pasteur". Over the past several decades, there has been signs that Williams and Nesse's idea are getting traction, and there is also growing evidence for their viewpoint that natural selection can lead to greater prevalence of gene variants are detrimental to health (Greaves, 2007; Pennisi, 2011). However, Darwin's Pangenesis has not been incorporated into the notion of Darwinian medicine yet. It is to be hoped that the concept of Darwinian medicine can be expanded with inclusion of Darwin's Pangenesis.

REFERENCES

Aaronson, S., & Todaro, S. (1969). Human diploid cell transformation by DNA extracted from the tumor virus SV40. *Science, 166*, 390–391.

Agbor-Enoh, S., Chan, J. L., Singh, A., Tunc, I., Gorham, S., Zhu, J., ... Valantine, H. (2018). Circulating cell-free DNA as a biomarker of tissue injury: Assessment in a cardiac xenotransplantation model. *The Journal of Heart and Lung Transplantation* (in press) https://doi.org/10.1016/j.healun.2018.04.009.

Aitken, R. (1917). A case of congenital transposition of the heart. *The British Medical Journal, 2*, 425.

Allan, G. A. (1910). A case of complete transposition of the viscera; with post-mortem report. *The British Medical Journal, 1*, 987.

Alvarado, A. S. (2009). A cellular view of regeneration. *Nature, 460*, 39–40.

Anderson, J. F. (1926). On Total Transposition of the Viscera and its Clinical Importance. *Journal of the Royal Society of Medicine, 19*, 119–120.

Arge, E. (1960). Transposition of the viscera and sterility in men. *Lancet, 275*, 412–414.

Aubertin, K., Silva, A. K. A., Luciani, N., Espinosa, A., Djemat, A., Charue, D., ... Wilhelm, C. (2016). Massive release of extracellular vesicles from cancer cells after photodynamic treatment or chemotherapy. *Scientific Reports, 6*, 35376.

Austin, L., Moses, E. K., & Langford, C. J. (1983). Macromolecular changes during nerve regeneration: Changes in small molecular weight RNAs. In B. Haber, J. R. Perez-Polo, G. H. Hashim, & A. M. Giuffrida-Stella (Eds.), *Nervous system regeneration* (pp. 471–477). New York: Liss.

Avery, O. T., MacLeod, C. M., & McCarty, M. (1944). Studies of the chemical nature of the substance inducing transformation of pneumococcal types. Induction of transformation by a desoxyribonucleic acid fraction isolated from *Pneumococcus* Type III. *Journal of Experimental Medicine, 79*, 137–158.

Barker, L. F. (1922). The relation of the endocrine glands to heredity and development. *Science, 55*, 685–690.

Basolo, F., Elliott, J., Tait, L., Chen, X. Q., Maloney, T., Russo, I. H., ... Russo, J. (1991). Transformation of human breast epithelial cells by c-Ha-ras oncogene. *Molecular Carcinogenesis, 4*, 25–35.

Bateson, W. (1913). *Problems of genetics*. New Haven: Yale University Press.

Beatty, J. (1997). Why Do Biologists Argue Like They Do? *Philosophy of Science, 64*, S432–S443.

Beck, J., Bierau, S., Balzer, S., Andag, R., Kanzow, P., Schmitz, J., ... Schutz, E. (2013). Digital droplet PCR for rapid quantification of donor DNA in the circulation of transplant recipients as a potential universal biomarker of graft injury. *Clinical Chemistry, 59*, 1732–1741.

Becker, A., Thakur, B. K., Weiss, J. M., Kim, H. S., Peinado, H., & Lyden, D. (2016). Extracellular vesicles in cancer: Cell-to-cell mediators of metastasis. *Cancer Cell, 30*, 836–848.

Bendich, A. (1961). Nucleic acids and the genesis of cancer. *Bulletin of the New York Academy of Medicine, 37*, 661–674.

Bendich, A., Borenfreund, E., Honda, Y., & Steinglass. (1969). Cell transformation and genesis of Cancer. *Archives of Environmental Health, 19*, 157–166.

Bendich, A., Wilczok, T., & Borenfreund, E. (1965). Circulating DNA as a possible factor in oncogenesis. *Science, 148*, 374–376.

Bergsmedh, A., Szeles, A., Henriksson, M., Bratt, A., Folkman, M. J., Spetz, A. L., & Holmgren, L. (2001). Horizontal transfer of oncogenes by uptake of apoptotic bodies. *Proceedings of the National Academy of Sciences of the United States of America, 98*, 6407–6411.

Bernstein, S. C., & Weinberg, R. A. (1985). Expression of the metastatic phenotype in cells transfected with human metastatic tumor DNA. *Proceedings of the National Academy of Sciences of the United States of America, 82*, 1726–1730.

Bisset, L. R., Boni, J., Lutz, H., & Schupbach, J. (2007). Lack of evidence for PERV expression after apoptosis-mediated horizontal gene transfer between porcine and human cells. *Xenotransplantation, 14*, 13–24.

Blegen, H. M. (1949). Surgery in situs in versus. *Annals of Surgery, 129*, 244–259.

Bodian, D. (1947). Nucleic acid in nerve-cell regeneration. *Symposia of the Society for Experimental Biology, 1*, 163–178.

Bracamonte-Baran, W., Florentin, J., Zhou, Y., Jankowska-Gan, E., Haynes, W. J., Zhong, W., ... Burlingham, W. J. (2017). Modification of host dendritic cells by microchimerism-derived extracellular vesicles generates split tolerance. *Proceedings of the National Academy of Sciences of the United States of America, 114*, 1099–1104.

Brassett, C., & Ellis, H. (1991). Transposition of the viscera: A review. *Clinical Anatomy, 4*, 139–147.

Burghoff, S., Ding, Z., Gödecke, S., Assmann, A., Wirrwar, A., Buchholz, D., ... Schrader, J. (2008). Horizontal gene transfer from human endothelial cells to rat cardiomyocytes after intracoronary transplantation. *Cardiovascular Research, 77*, 534–543.

Bynum, W. F. (1983). Darwin and the doctors: Evolution, diathesis, and germs in 19th-century Britain. *Gesnerus, 40*, 43–53.

Cai, J., Wu, G., & Jose, P. A. (2016). Functional transferred DNA within extracellular vesicles. *Experimental Cell Research, 349*, 179–183.

Campbell, M. (1963). The mode of inheritance in isolated laevocardia, dextrocardia and situs inversus. *British Heart Journal, 25*, 803–813.

Chan, K. C., Woo, J. K., King, A., Zee, B. C. Y., Lam, W. K. J., Chan, S. L., ... Lo, Y. M. D. (2017). Analysis of plasma Epstein–Barr virus DNA to screen for nasopharyngeal cancer. *New England Journal of Medicine, 377*, 513–522.

Chen, Q., Yan, W., & Duan, E. K. (2016). Epigenetic inheritance of acquired traits through sperm RNAs and sperm RNA modifications. *Nature Reviews Genetics, 7*, 733–743.

Cockayne, E. A. (1938). The Genetics of transposition of the viscera. *Quarterly Journal of Medicine, 7*, 479–493.

Darwin, C. (1868). *The variation of animals and plants under domestication*. London: John Murray.

Darwin, C. (1875). *The variation of animals and plants under domestication* (2nd ed.). London: John Murray.

Darwin, C. (1881). Inheritance. *Nature, 24*, 257.

Darwin, F. (1888). *The life and letters of Charles Darwin*. London: John Murray.

Di Liegro, C. M., Schiera, G., & Di Liegro, I. (2017). Extracellular vesicle-associated RNA as a Carrier of epigenetic information. *Genes, 8*, 240.

Egan, S. E., Wright, J. A., Jarolim, L., Yanagihara, K., Bassin, R. H., & Greenberg, A. H. (1987). Transformation by oncogenes encoding protein kinases induces the metastatic phenotype. *Science, 238*, 202–205.

Eulalio, A., Mano, M., Dal Ferro, M., Zentilin, L., Sinagra, G., Zacchigna, S., & Giacca, M. (2012). Functional screening identifies miRNAs inducing cardiac regeneration. *Nature, 492*, 376–381.

Fidler, I. J. (2003). The pathogenesis of cancer metastasis: The 'seed and soil' hypothesis revisited. *Nature Reviews Cancer, 3*, 1–6.

Foerster, A. (1861). *Die Missbildung des Menschen*. Jena. p136.

Gala-Lopez, B. L., Neiman, D., Kin, T., O'Gorman, D., Pepper, A. R., Malcolm, A. J., ... Shapiro, A. M. J. (2018). Beta cell death by cell-free DNA and outcome after clinical islet transplantation. *Transplantation, 102*, 978–985.

Gannett, W. W. (1889). A case of transposition of all the organs. *Boston Medical & Surgical Journal, 120*, 104.

Garcia-Olmo, D. C., Domínguez, C., García-Arranz, M., Anker, P., Stroun, M., Garcia-Verdugo, J. M., & Garcia-Olmo, D. (2010). Cell-free nucleic acids circulating in the plasma of colorectal cancer patients induce the oncogenic transformation of susceptible cultured cells. *Cancer Research, 70*, 560–567.

Garcia-Olmo, D., Garcia-Olmo, D. C., Ontanon, J., Martinez, E., & Vallejo, M. (1999). Tumor DNA circulating in the plasma might play a role in metastasis — the hypothesis of the genomemetastasis. *Histology & Histopathology, 14*, 1159–1164.

García-Olmo, D. C., Samos, J., Picazo, M. G., Asensio, A. I., Toboso, I., & García-Olmo, D. (2008). Release of cell-free DNA into the bloodstream leads to high levels of non- tumor plasma DNA during tumor progression in rats. *Cancer Letter, 272*, 133–140.

Gielis, E. M., Ledegance, K. J., de Winter, B. Y., Del Favero, J., Bosmans, J. L., Claas, F. H., ... Eikmans, M. (2015). Cell-free DNA: An upcoming biomarker in transplantation. *American Journal of Transplantation, 15*, 2541–2545.

Goh, S. K., Muralidharan, V., Christophi, C., Do, H., & Dobrovic, A. (2017). Probe- free droplet digital PCR to measure donor specific cell-free DNA after solid-organ transplantation. *Clinical Chemistry, 63*, 742–750.

Gonzalez-Nolasco, B., Wang, M., Prunevieille, A., & Benichou, G. (2018). Emerging role of exosomes in allorecognition and allograft rejection. *Current Opinion in Organ Transplantation, 23*, 22–27.

Greaves, M. (2007). Darwinian medicine: a case for cancer. *Nature Reviews Cancer, 7*, 213–221.

Grskovic, M., Hiller, D. J., Eubank, L. A., Sninsky, J. J., Christopherson, C., Collins, J. P., ... Woodward, R. N. (2016). Validation of a clinical-grade assay to measure donor-derived cell-free DNA in solid organ transplant recipients. *Journal of Molecular Diagnostics, 18*, 890–902.

Hackam, A. S. (2000). Breaking the law: The Molecular Basis of Non-Mendelian Transmission. *Clinical Genetics, 57*, 173–175.

Hasek, M. (2003). Vegetative hybridization of animals by joint blood circulation during embroyonal development. *Transplantation, 76*, 1412–1421.

Hayashi, T., & Hoffman, M. P. (2017). Exosomal microRNA communication between tissues during organogenesis. *RNA Biology, 14*, 1683–1689.

Higami, Y., & Shimokawa, I. (2000). Apoptosis in the aging process. *Cell and Tissue Research, 301*, 125–132.

Hodgkinson, C. P., Kang, M. H., Dal-Pra, S., Mirotsou, M., & Dzau, V. J. (2015). MicroRNAs and cardiac regeneration. *Circulation Research, 116*, 1700–1711.
Holm, M. M., Kaiser, J., & Schwab, M. E. (2018). Extracellular vesicles: Multimodel envoys in neural maintenance and repair. *Trends in Neurosciences, 41*, 360–372.
Horsfall, F. L. (1962). Heritance of acquired characteristics. *Science, 136*, 472–476.
Hur, J. Y., Kim, H. J., Lee, J. S., Choi, C.-M., Lee, J. C., Jung, M. K., ... Lee, K. Y. (2018). Extracellular vesicle-derived DNA for performing EGFR genotyping of NSCLC patients. *Molecular Cancer, 17*, 15.
Hutcheson, J. D., & Aikawa, E. (2018). Extracellular vesicles in cardiovascular homeostasis and disease. *Current Opinion in Cardiology, 33*, 290–297.
Janowska-Wieczorek, A., Wysoczynski, M., Kijowski, J., Marquez-Curtis, L., Machalinski, B., Ratajczak, I., & Rataiczak, M. Z. (2005). Microvesicles derived from activated platelets induce metastasis and angiogenesis in lung cancer. *International Journal of Cancer, 113*, 752–760.
Jayawardena, T. M., Egemnazarov, B., Finch, E. A., Zhang, L., Payne, J. A., Pandya, K., ... Dzau, V. J. (2012). MicroRNA-mediated in vitro and in vivo direct reprogramming of cardiac fibroblasts to cardiomyocytes. *Circulation Research, 110*, 1465–1473.
Jirtle, R. L., & Skinner, M. K. (2007). Environmental epigenomics and disease susceptibility. *Nature Reviews Genetics, 8*, 253–262.
Jylhava, J., Kotipelto, T., Raitala, A., Jylha, M., Hervonen, A., & Hurme, M. (2011). Aging is associated with quantitative and qualitative changes in circulating cell-free DNA: The vitality 90 + study. *Mechanisms of Ageing and Development, 132*, 20–26.
Kahlert, C., Melo, S. A., Protopopov, A., Tang, J., Seth, S., Koch, M., ... Kalluri, R. (2014). Identification of double-stranded genomic DNA spanning all chromosomes with mutated KRAS and p53 DNA in the serum exosomes of patients with pancreatic cancer. *The Journal of Biological Chemistry, 289*, 3869–3875.
Kalluri, R., & LeBleu, V. S. (2016). .Discovery of double-stranded genomic DNA in circulating exosomes. *Cold Spring Harbor Symposia on Quantitative Biology, 81*, 275–280.
Kapur, N. C. (1924). Complete transposition of viscera. *Lancet, 204*, 1097.
Kawamura, Y., Yamamoto, Y., Sato, T. A., & Ochiya, T. (2017). Extracellular vesicles as trans-genomic agents: Emerging roles in disease and evolution. *Cancer Science, 108*, 824–830.
Ke, Y., Beesley, C., Smith, P., Barraclough, R., Rudland, P. S., & Foster, C. S. (1998). Generation of metastatic variants by transfection of a rat non-metastatic epithelial cell line with genomic DNA from rat prostatic carcinoma cells. *British Journal of Cancer, 77*, 287–296.
Keshtkar, S., Azarpira, N., & Ghahremani, M. H. (2018). Mesenchymal stem cell-derived extracellular vesicles: Novel frontiers in regenerative medicine. *Stem Cell Research & Therapy, 9*, 63.
Klein, B. Y. (1981). A suggested mechanism for changing tumor cell phenotype: Transfection of host cells with DNA sequences of dead tumor cells. *Medical Hypotheses, 7*, 645–650.
Knudson, A. G. (1971). Mutation and cancer: Statistical study of retinoblastoma. *Proceedings of the National Academy of Sciences of the United States of America, 68*, 820–823.
Kong, W., Nuo, M., Zhu, X. P., Han, X. J., & Wang, X. (2013). Kidney regeneration by nonplatelet RNA-containing particle-derived cells. *Clinical and Experimental Pharmacology & Physiology, 40*, 724–734.
Kristoffersson, U., & Macek, M. (2017). From Mendel to Medical genetics. *European Journal of Human Genetics, 25*, S53–S59.

Kushner, H. F. (1957). The influence of metabolic factors on the heredity of animals. *Proceedings of the International Genetics Symposia, Tokyo and Kyoto, Japan*, 515–519.

Lee, T. H., Chennakrishnaiah, S., Audemard, E., Montermini, L., Meehan, B., & Rak, J. (2014). Oncogenic ras-driven cancer cell vesiculation leads to emission of double-stranded DNA capable of interacting with target cells. *Biochemical and Biophysical Research Communications, 451*, 295–301.

Lee, T. H., Chennakrishnaiah, S., Meehan, B., Montermini, L., Garnier, D., D'Asti, E., … Rak, J. (2016). Barriers to horizontal cell transformation by extracellular vesicles containing oncogenic H- ras. *Oncotarget, 7*, 8–11.

Leininger, C. R., & Gibson, S. (1950). Transposition of viscera in siblings. *The Journal of Pediatrics, 37*, 195–200.

Leon, S. A., Shapiro, B., Sklaroff, D. M., & Yaros, M. J. (1977). Free DNA in the serum of cancer patients and the effect of therapy. *Cancer Research, 37*, 646–650.

Lesizza, P., Prosdocimo, G., Martinelli, V., Sinagra, G., Zacchigna, S., & Giacca, M. (2017). Single-dose intracardiac injection of pro-regenerative microRNAs Improves cardiac function after myocardial infarction. *Circulation Research, 120*, 1298–1304.

Levin, B. R., Lipsitch, M., & Bonhoeffer, S. (1999). Population biology, evolution, and infectious disease: Convergence and synthesis. *Science, 283*, 806–809.

Lia, G., Brunello, L., Bruno, S., Carpanetto, A., Omedè, P., Festuccia, M., … Bruno, B. (2018). Extracellular vesicles as potential biomarker of acute Graft-versus-host- disease. *Leukemia, 32*, 765–773.

Lichtenstein, P., Holm, N. V., Verkasalo, P. K., Iliadou, A., Kappio, J., Koskenvuo, M., … Hemminki, K. (2000). Environmental and heritable factors in the causation of cancer — analyses of cohorts of twins from Sweden, Denmark, and Finland. *New England Journal of Medicine, 343*, 78–85.

Lichtenstein, A. V., Melkonyan, H. S., Tomei, L. D., & Umansky, S. R. (2001). Circulating nucleic acids and apoptosis. *Annals of the New York Academy of Sciences, 945*, 239–249.

Liddon, E. (1904). Note on a case of transposition of the viscera. *Lancet, 163*, 1197.

Lisi, V., Singh, B., Giroux, M., Guzman, E., Painter, M. W., Cheng, Y. C., … Kosik, K. S. (2017). Enhanced neuronal regeneration in the CAST/Ei mouse strain is linked to expression of differentiation markers after injury. *Cell Reports, 20*, 1136–1147.

Liu, Y.-S. (2006). Historical and modern genetics of plant graft hybridization. *Advances in Genetics, 56*, 101–129.

Liu, Y.-S. (2008). A new perspective on Darwin's Pangenesis. *Biological Reviews of the Cambridge Philosophical Society, 83*, 141–149.

Liu, Y.-S. (2010). Darwin's gemmules and oncogenes. *Annals of Oncology, 21*, 908.

Liu, Y.-S., & Li, X.-J. (2012). Darwińs Pangenesis and molecular medicine. *Trends in Molecular Medicine, 18*, 506–508.

Liu, Y., & Li, X. (2016). Darwin's Pangenesis as a molecular theory of inherited diseases. *Gene, 582*, 19–22.

Lo, Y. M. D., Tein, M. S., Pang, C. C., Yeung, C. K., Tong, K. L., & Hjelm, N. M. (1998). Presence of donor-specific DNA in plasma of kidney and liver transplant recipients. *Lancet, 351*, 1329–1330.

Mahar, M., & Cavalli, V. (2018). Intrinsic mechanisms of neuronal axon regeneration. *Nature Reviews Neuroscience, 19*, 323–337.

Méthot, P.-O. (2015). Darwin, evolution, and medicine: Historical and contemporary perspectives. In T. Heams, P. Huneman, G. Lecointre, & M. Silberstein (Eds.), *Handbook of evolutionary thinking in the sciences* (pp. 587–617). Netherlands: Springer.

Mittelbrunn, M., & Sanchez-Madrid, F. (2012). Intercellular communication: Diverse structures for exchange of genetic information. *Nature Reviews Molecular Cell Biology, 13*, 328–335.

Neofytou, M., Brison, N., Van den Bogaert, K., Dehaspe, L., Devriendt, K., Geerts, A., & Vermeesch, J. R. (2018). Maternal liver transplant: Another cause of discordant fetal sex determination using cell-free DNA. *Prenatal Diagnosis, 38*, 148–150.
Newman, E. A., & Grossman, M. I. (1951). Effect of nucleic acid supplements in the diet on rate of regeneration of liver in rats. *American Journal of Physiology, 164*, 251–253.
Ostrer, H. (1998). *Non-Mendelian Genetics in Humans*. New York: Oxford University Press.
Paget, S. (1889). The distribution of secondary growths in cancer of the breast. *Lancet, 133*, 571–573.
Parsonnet, A. E. (1929). A case of complete transposition of the viscera with electrocardiographic and X-ray studies. *Annals of Internal Medicine, 2*, 963–967.
Pennisi, E. (2011). Evolution: Darwinian medicine's drawn-out dawn. *Science, 334*, 1486–1487.
Petronis, A. (2010). Epigenetics as a unifying principle in the aetiology of complex traits and diseases. *Nature, 465*, 721–727.
Plaks, V., Koopman, C. D., & Werb, Z. (2013). Circulating tumor cells. *Science, 341*, 1186–1188.
Risch, N. J. (2000). Searching for genetic determinants in the new millennium. *Nature, 405*, 847–856.
Risdon, W. E. (1906). Transposition of viscera. *The British Medical Journal*, 455–455.
Ross, J. (1872). *The graft theory of disease, being an application of Mr. Darwin's hypothesis of Pangenesis to the explanation of the phenomena of the phenomena of the Zymotic diseases*. Philadelphia: Lindsay and Blakiston.
Schutz, E., Fischer, A., Beck, J., Harden, M., Koch, M., Wuensch, T., … Oellerich, M. (2017). Graft-derived cell-free DNA, a noninvasive early rejection and graft damage marker in liver transplantation: A prospective, observational, multicenter cohort study. *PLoS Medicine, 14*, e1002286.
Schwarzenbach, H., Hoon, D. S. B., & Pantel, K. (2011). Cell-free nucleic acids as biomarkers in cancer patients. *Nature Reviews Cancer, 11*, 426–437.
Sharon, E., Shi, H., Kharbanda, S., Koh, W., Martin, L. R., Khush, K. K., … de Vlaminck, I. (2017). Quantification of transplant-derived circulating cell-free DNA in absence of a donor genotype. *PLoS Computational Biology, 13*, e1005629.
Sherman, S. L. (1997). Evolving Methods in Genetic Epidemiology. IV. Approaches to Non-Mendelian Inheritance. *Epidemiologic Reviews, 19*, 44–51.
Shih, C., Padhy, L. C., & Weinberg, R. A. (1981). .Transforming genes of carcinomas and neuroblastomas introduced into mouse fibroblasts. *Nature, 290*, 261.
Slavkin, H. C. (1997). Charles Darwin and the foundations of clinical genetics in dentistry. *Journal of the American Dental Association, 128*, 241–245.
Stewart, C. M., Kothari, P. D., Mouliere, F., Mair, R., Somnay, S., Benayed, R., … Tsui, D. W. Y. (2018). The value of cell-free DNA for molecular pathology. *The Journal of Pathology, 244*, 616–627.
Stroun, M., Anker, P., Maurice, P., Lyautey, J., Lederrey, C., & Beljanski, M. (1989). Neoplastic characteristics of the DNA found in the plasma of cancer patients. *Oncology, 46*, 318–322.
Thatcher, E. J., & Patton, J. G. (2010). Small RNAs have a big impact on regeneration. *RNA Biology, 7*, 33–38.
Themeli, M., Waterhouse, M., Finkle, J., & Spyridonidis, A. (2011). DNA chimerism and its consequences after allogenic hematopoietic cell transplantation. *Chimerism, 2*, 25–28.
Thierry, A. R., El Messaoudi, S., Gahan, P. B., Anker, P., & Stroun, M. (2016). Origins, structures, and functions of circulating DNA in oncology. *Cancer Metastasis Reviews, 35*, 347–376.

Thorgeirsson, U. P., Turpeenniemi-Hujanen, T., Williams, J. E., Westin, E. H., Heilman, C. A., Talmadge, J. E., & Liotta, L. A. (1985). NIH 3T3 cells transfected with human tumor DNA containing activated ras oncogenes express the metastatic phenotype in nude mice. *Molecular and Cellular Biology, 5*, 259–262.

Tihen, J. A., Charles, D. R., & Sippel, T. O. (1948). Inherited visceral inversion in mice. *Journal of Heredity, 39*, 29–31.

Torgersen, J. (1946). Familial transposition of viscera. *Acta Medica Scandinavica, 126*, 319–322.

Towers, B. (1968). The impact of Darwin's Origin of Species on medicine and biology. In F. N. L. Poynter (Ed.), *Medicine and science in the 1860s* (p. 49). London: Wellcome Institute of the History of Medicine.

Trejo-Becerril, C., Pérez-Cárdenas, E., & Dueñas-González, A. (2014). In vivo rat model to study horizontal tumor progression. *Methods in Molecular Biology, 1165*, 175–185.

Trejo-Becerril, C., Pérez-Cárdenas, E., Taja-Chayeb, L., Anker, P., Herrera-Goepfert, R., Medina-Velázquez, L. A., … Dueñas-González, A. (2012). Cancer progression mediated by horizontal gene transfer in an in vivo model. *PLoS One, 7*, e52754.

Valadi, H., Ekstrom, K., Bossios, A., Sjostrand, M., Lee, J. J., & Lotvall, J. O. (2007). Exosome-mediated transfer of mRNAs and microRNAs is a novel mechanism of genetic exchange between cells. *Nature Cell Biology, 9*, 654–659.

Vanharanta, S., & Massague, J. (2013). Origins of metastatic traits. *Cancer Cell, 24*, 410–421.

Wan, J. C., Massie, C., Garcia-Corbacho, J., Mouliere, F., Brenton, J. D., Caldas, C., … Rosenfeld, N. (2017). Liquid biopsies come of age: Towards implementation of circulating tumour DNA. *Nature Reviews Cancer, 17*, 223–238.

Watanabe, K., Yamaji, R., & Ohtsuki, T. (2018). MicroRNA-664a-5p promotes neuronal differentiation of SH-SY5Y cells. *Genes to Cells, 23* (in press) https://doi.org/10.1111/gtc.12559.

Waterhouse, M., Themeli, M., Bertz, H., Zoumbos, N., Finke, J., & Spyridonidis, A. (2011). Horizontal DNA transfer from donor to host cells as an alternative mechanism of epithelial chimerism after allogeneic hematopoietic cell transplantation. *Biology of Blood and Marrow Transplantation, 17*, 319–329.

Williams, G. C., & Nesse, R. M. (1991). The dawn of Darwinian medicine. *The Quarterly Review of Biology, 66*, 1–22.

Williamson, M. B., & Guschlbauer, W. (1961). Metabolism of nucleic acids during regeneration of wound tissue. *The Journal of Biological Chemistry, 236*, 1463–1466.

Williamson, M. B., & Guschlbauer, W. (1963). Metabolism of nucleic acids during regeneration of wound tissue. II. The rate of formation of RNA. *Archives of Biochemistry and Biophysics, 100*, 245–250.

Winther, R. G. (2000). Darwin on the variation and heredity. *Journal of the History of Biology, 33*, 425–455.

Wu, L. E., Gomes, A. P., & Sinclair, D. A. (2014). Geroncogenesis: Metabolic changes during aging as a driver of tumorigenesis. *Cancer Cell, 25*, 12–18.

Zangi, L., & Hajjar, R. J. (2017). Synthetic microRNAs stimulate cardiac repair. *Circulation Research, 120*, 1222–1223.

Zhang, J., Tong, K. L., Li, P. K., Chan, A. Y., Yeung, C. K., Pang, C. C., … Lo, Y. M. D. (1999). Presence of donor- and recipient derived DNA in cell-free urine samples of renal transplantation recipients: Urinary DNA chimerism. *Clinical Chemistry, 45*, 1741–1746.

CHAPTER FOUR

Darwin's Pangenesis and Certain Anomalous Phenomena

Yongsheng Liu*,§
*Henan Collaborative Innovation Center of Modern Biological Breeding, Henan Institute of Science and Technology, Xinxiang, China
§Department of Biochemistry, University of Alberta, Edmonton, AB, Canada
E-mail: ysliu63@yahoo.ca.

Contents

Abstract

Darwin clearly described certain anomalous phenomena, including what he referred to as "the direct action of the male element on the female form" and what we now call xenia and telegony, bud variation (mutation), reversion or atavism, and the inheritance and non-inheritance of mutilation. Some phenomena, particularly xenia, telegony and the inheritance of mutilation, were considered as doubtful phenomena by such authorities as Weismann and Morgan. Over the past 150 year, however, there has been increasing evidence for xenia, which is of great interest and importance in physiological research and plant production. The discoveries of cell-free fetal DNA,

Advances in Genetics, Volume 102
ISSN 0065-2660
https://doi.org/10.1016/bs.adgen.2018.05.009

sperm RNAs, penetration of sperm into the somatic tissues of the female reproductive tract and the incorporation of exogenous DNA into somatic cells indicate that molecular mechanisms exist for telegony, one of the most controversial issues. Darwin's Pangenesis is the only theory that explains all the different types of phenomena.

1. XENIA: EFFECTS OF POLLEN ON SEEDS AND FRUITS

1.1 Darwin on Xenia

Xenia is described as the direct or immediate pollen effect on the size, shape, colour, developmental timing and chemical composition of seeds and fruits (Denney, 1992). Before Focke (1881, pp. 510–518) coined the term "xenia", Darwin (1868) collected many cases, and devoted six pages to what he referred to as "the direct action of the male element on the female form". He believed that this was one remarkable fact, and had a "high physiological importance". He first described the direct action of pollen in peas, and pointed out that, as early as 1729, it was observed that when white and blue varieties planted near each other, they crossed, and in the autumn blue and white peas were found within the same pods. He cited Gartner, who "selected the most constant varieties, and the result conclusively showed that the colour of the skin of the pea is modified when pollen of a differently coloured variety is used." He also cited Laxton, who fertilized the Tail Sugar-pea, which bore very thin green pods, with pollen of the Purple-podded pert, which, had dark-purple pods with very thick skin. The result showed that "a flower fertilized by pollen from the purple-pod yielded a pod clouded with purplish-red." This observation was confirmed by Darwin's own experiments.

According to Darwin (1875), as long ago as 1751, it was observed that, when differently-coloured varieties of maize grew near each other, they mutually affected each other's seeds. He described the experiments performed by Savi, who "sowed yellow and black-seeded maize together, and on the same ear some of the seeds were yellow, some black, and some mottled, the differently coloured seeds being arranged irregularly or in rows." Darwin also cited an interesting experiments conducted by Arnold, who showed that "a female flower was subjected first to the action of pollen from a yellow variety, and then to that from a white variety; the result was an ear, each grain of which was yellow below and white above." Addition to maize, Darwin also described similar pollen effects in other annual plants.

With regards to fruit trees, Darwin (1875) cited Gallesio, who "fertilized the flowers of an orange with pollen from the lemon; and one fruit thus produced bore a longitudinal stripe of peel having the colour, flavour, and other characters of the lemon." He claimed that "no case of the direct action of the pollen of one variety on another is better authenticated or more remarkable than that of the common apple." He gave several examples: "in one of these a Russeting apple and an adjoining kind mutually affected each other's fruit; and in another case a smooth apple affected a rough-coated kind." He also noted that "two very different apple-trees growing close to each other, which bore fruit resembling each other, but only on the adjoining branches." After giving the examples, Darwin (1875) concluded as follows:

I have now shown, on the authority of several excellent observers, in the case of plants belonging to widely different orders, that the pollen of one species or variety, when applied to the female of a distinct form, occasionally causes the coats of the seeds, the ovarium or fruit, including even the calyx and upper part of the peduncle of the apple, and the axis of the ear in maize, to be modified. Sometimes the whole ovarium or all the seeds are thus affected; sometimes only a certain number of the seeds, as the case of the pea, or only a part of the ovarium, as with the striped orange, mottled grapes, and maize, is thus affected. ... The proofs of the action of foreign pollen on the mother-plant have been given in considerable detail, because this action, as we shall see in a future chapter, is of the highest theoretical importance, and because it is in itself a remarkable and apparently anomalous circumstance. That it is remarkable under a physiological point of view is clear, for the male element not only affects, in accordance with its proper function, the germ, but at the same time various parts of the mother-plant, in the same manner, as it affects the same part in the seminal offspring from the same two parents. We thus learn that an ovule is not indispensable for the reception of the influence of the male element. But this direct action of the male element is not so anomalous as it at first appears, for it comes into play in the ordinary fertilisation of many flowers .... Therefore, in the previously given cases we have only to believe in the further power of pollen, when applied to a distinct species or variety, to influence the shape, size, colour, texture, etc., of certain parts of the mother plant.

The phenomena of "the direct action of pollen on the mother-plant" and graft hybridization were essential to Darwin's thinking when he formulated his Pangenesis. In a letter, Darwin wrote: "You will see why I think these two subjects so important. They have led me to form a hypothesis on the various forms of reproduction, development, inheritance, etc., which hypothesis, I believe, will ultimately be accepted, though how it will be now received I am very doubtful." (Darwin, 1888) According to Darwin's Pangenesis, the formative matter within the pollen of plants, which consists of gemmules, can unite with and modify the partially developed cells of the mother plant. He believed that "the gemmules derived from the foreign pollen do not become developed merely in contact with pre-existing cells, but actually penetrate the nascent cells of the mother plant. According to this view, the cells of the mother-plant may almost literally be said to be fertilized by the gemmules derived from the foreign pollen."

1.2 Further Evidence for Xenia

Over the past 150 years, there has been increasing evidence for xenia phenomenon, which has applications not only in genetics and physiological research but also in plant breeding and crop production. In 1992, Denney collected a range of experimental and observational data and published a comprehensive review articles on xenia. This article was highly appreciated and included many interesting study results. He listed 3 tables and included many plants involved, such as apple, bean, blueberry, chestnut, cotton, coconut, corn, data, grape, olive, pea, pear, peach, pecan, persimmon, pistachio, pomegranate, radish, sunflower, sapota, sweet cherry, stock and so on (Denney, 1992). To avoid duplication, I would like to list the references that published only after 1992 (see Table 1). According to incomplete statistics, there are nearly 300 papers on xenia in the scientific literature. Some papers were published in *Science*, *Proceedings of the National Academy of Sciences of the United States of America,* and other prestigious journals (McKay & Crane, 1939; Nebel & Trump, 1932). Now it seems that xenia is no longer one of the doubtful phenomena, as Weismann (1912) proposed.

Since Focke (1881, pp. 510–518) used the term xenia with reference to pollen effect on endosperm of maize, this plant has become the classical example. Several experiments have shown an effect of pollen on kernel weight and yield. Kiesselbach and Leonard (1932) reported that compared to self fertilization, cross fertilization increased kernel weights on average by 10.1%, and showed an increase in yield of 35% in the sweet/dent cross in comparison with sweet/sweet with the same female. Bulant and Gallais (1998) observed the same phenomenon and showed that when inbred lines were used as females the average increase in kernel weight at maturity, because of fertilization, was about 12%. It should be noted that when a plant is male-sterile, an open pollination by a pollinator is necessary. Weingartner, Kaeser, Long, and Stamp (2002) reported that the best combination of a hybrid with cytoplasmic male sterility (CMS) and a non-isogenic pollinator hybrid increased the grain yield by 21.4% compared with the male-fertile isogenic counterpart of the CMS hybrid. This is of great economic importance in maize production. In addition, the xenia effects of pollen-induced changes on seed colour is also obvious (see Fig. 1).

Table 1 Evidence for Xenia Effects in Plants (After 1992, Incomplete Statistics)

Species	Parts and Characters Affected	References
Almond	amygdalin content	Sanchez-Perez, Arrazola, Martin, Grane, and Dicenta (2012)
	oil content/fatty acid	Kodad, Estopanan, Juan, and Company (2009)
Apple	fruit size/shape	Lukic, Maric, Milosevic, and Mitrovic (2015)
Apricot	Fruit size/colour	Yang, Yao, Zhou, and Yang (2002).
Bean	seed weight	Duc, Moessner, Moussy, and Mousset-Declas (2001)
Cotton	embryo weight	Pahlavani and Abolhasani (2006)
Date palm	fruit set and yield	Mohammadi et al. (2017); Salomon-Torresl et al. (2017)
Grape	berry/seed characters	Sabir (2015)
Hazelnut	nut/kernel characters	Fattahi, Mohammadzedeh, and Khadivi-Khub (2014)
Kiwifruit	fruit colour	Seal, Dunn, de Silva, McGhie, and Lunken (2013)
Litchi	fruit quality	Qiu et al. (2006)
Maize	kernel weight	Bulant and Gallais (1998).
	Grain yield	Weingartner et al. (2002); Weingartner, Camp, and Stamp (2004)
	Oil content	Duan, Song, Wang, Fan, and Zhao (2002)
Macadamia	nut/kernel weight	Wallace et al. (1996)
Mandarin	fruit set/sugar content	Wallace and Lee (1999)
Paeonia ostii	seed oil	Xie et al. (2017)
Pomegranate	fruit quality/seed characteristics	Gharaqhani et al. (2017)
Tomato	fruit/seed weight	Piotto, Bataqin-Piotto, de Almeida, and Oliveira (2013)

There is increasing experimental evidence for xenia in date palm. It has been found very striking influence of pollen, not only on the size, shape and other characters of the seed and fruit, but also on the time of ripening (Nixon, 1928, 1934; Swingle, 1928). Recently, these observations have been fully confirmed by many researchers in different countries (Gupta, Godara, Panda, Sharma, & Kaushik, 2017; Mohammadi, Rastgoo, & Izadi, 2017; Salomon-Torresl et al., 2017). For example Salomon-Torresl et al. (2017) pollinated the variety "Medjool" with the pollen from four different varieties ("Medjool", "Zahidi", "Khadrawy", and "Deglet Noor"). The results demonstrated that bunches pollinized with "Zahidi" had the highest

Figure 1 Xenia phenomenon illustrated by ears that were fertilized with mixtures of pollen from the same and another variety. Ears of two on left are from yellow dent corn, two on right are from corn variety with blue colour. Two center ears are from adjacent rows, and two outside ears are from plants approximately 40 rows away from blue or yellow kernel plants. *Reproduced with the permission from Peter Thomison, The Ohio State University.*

fruit set percentage (50.30%), whereas bunches pollinized with "Khadrawy" gave the lowest set percentage (13.25%). The highest total yield (64.55 kg) was obtained with "Zahidi" pollen, whereas the lowest total yield (41.79 kg) was obtained with "Deglet Noor" pollen.

There are many reports demonstrating the xenia effects in fruit trees, suggesting the need for careful consideration of interplanting of varieties and managements of pollinators to maximize fruit production and quality. For example Wallace and Lee (1999) investigated the effects of pollen on fruit set and fruit quality characters in three mandarin varieties ("Imperial", "Ellenor" and "Murcott"). They found that pollen source significantly affected fruit set, seed number and sugar content of "Ellenor". In particular, "Murcott" pollen produced a significantly higher fruit set, relatively low seed number, and the highest mean sugar content (13.2%), significantly higher than "Imperial" pollen, bagged and unpollinated treatments. Recently

Gharaqhani, Soloklui, Oraquzie, and Zare (2017), determined the effect of pollen source on fruit, aril, and seed characters in pomegranate. The variety "Malas Yazdi" was pollinated with three pollen sources ("Rabab Neyriz", "Malas Pishva Varamin", and "Poost Sefid Dezful"). The results demonstrated that the rate of fruit set from different pollen sources varied from 13% to 46%. Most recently, it has been demonstrated that a careful choice of compatible pollinizers significantly improves the size of fruits in red raspberry, suggesting that selection of suitable pollinizer is essential for high yield production (Zurawicz, Studnicki, Kubik, & Pruski, 2018).

In some plants, the xenia effects could be extremely significant. Freytag (1979) reported that normal maternal control of pod development in the bean line MITA 10597 was influenced in the rate of elongation and the total length of the mature pod when flowers were fertilized by foreign pollen. A plant that normally produces very small pods (MITA 10597) produced pods up to twice as large when the pod was fertilized by pollen from other bean lines (see Fig. 2).

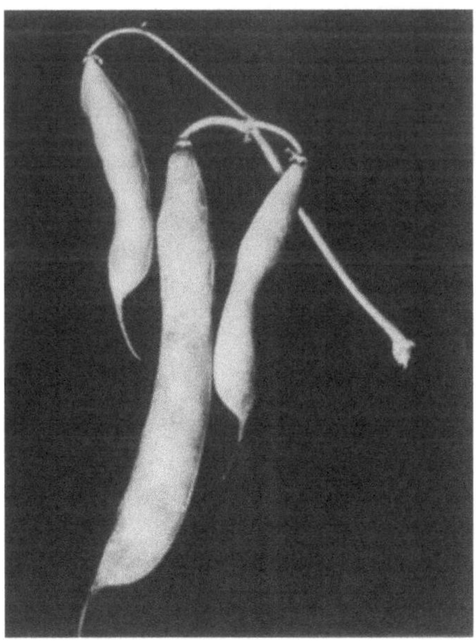

Figure 2 The metaxenia effects on bean pods. The long center pod received pollen from 2W-33-2 and the two smaller pods are from selfed flowers, on a single inflorescence of MITA 10597. *Reproduced with permission from Freytag, G.F. (1979). Metaxenia effects on pod size development in the common bean.* Journal of Heredity, 70, *444–446.*

Mizrahi, Mouyal, Nerd and Sitrit (2004) provided convincing evidence for the existence of xenia in the vine cacti, demonstrating that pollen source not only affected fruit size, pulp dry weight and number of seeds per fruit, but also affected the fruit-ripening time. By comparing with the ripening time of fruits from *Hylocereus polyrhizus* pollinated with *Hylocereus undatus* pollen, those pollinated with the pollen from *Selenicereus grandiflorus* and *S. megalanthus* delayed ripening by 1 and 3 weeks, respectively (see Fig. 9.3). This xenia effect on the fruit-ripening time can be used for expanding the marketing period.

Figure 3 The branch of *Hylocereus polyrhizus* was bearing two fruits at different ripening stages. The red, ripe fruit produced from pollination with the pollen of *H. undatus* clones, and the green, unripe fruit was from pollination with *S. grandiflorus*. This picture was taken 34 days after pollination. *Reproduced with permission from Mizrahi, Y., Mouyal, J., Nerd, A., & Sitrit, Y. (2004). Metaxenia in the vine cacti Hylocereus polyrhizus and Selenicereus spp. Annals of Botany, 93, 469–472.*

1.3 Mobile RNAs: A Novel Mechanism for Xenia?

Now the practical importance of xenia has been realized by many plant growers, but its mechanism remains poorly understood and awaits further research. de Vries and Correns, the two chief actors in the founding of genetics, also realized the importance of xenia, but their explanations for this phenomenon are different from Darwin's. They showed that xenia in maize could be explained by double fertilization (Dunn, 1973). Double fertilization may be the explanation of the direct effect of pollen on the seed, but it does not count for the influence on the fruit outside the embryo and endosperm.

Swingle (1928) surmised that there is an analogy to hormone action in the bodies of animals: "The simplest and most probable theory to explain metaxenia is that the embryo or endosperm or both of them secrete hormones, or soluble substances analogous to them, which diffuse out into the tissues of the mother plant that constitute the seed and fruit and there exert a specific effect on these tissues, varying according to the particular male parent used to fecundate the embryo and endosperm." Dag and Mizrahi (2005) suggested that seed number is correlated with fruit size in many crops. They supposed that the developing seeds secrete hormones (mainly auxins) that promote development of the ovary, leading to larger fruit. Although this hormone action hypothesis is widely accepted, it remains unsatisfactory because the parallelism in plants between the xenia effects and the characters of the pollen parent suggests that secretions may be initiated by the genes which determine the characters. In addition Denney (1992) asked a good question: if the embryo secretes hormones affecting maternal tissue, how can the endosperm and the embryo itself not be affected by these same hormones?

Transposons have also been considered as candidates for the root cause of many xenia phenomena. However, transposons only move within and between chromosomes. They do not "jump" from syngamous tissue to maternal tissue, as (Denney, 1992) pointed out. Kermicle (1978) described the phenomenon of imprinting, in which there is preferential expression of genes depending on whether the genes were inherited from the paternal or the maternal side. Paternal imprinting can clearly affect endosperm development and kernel size in corn, thus explains certain cases of xenia. Such imprinting, or parent-specific gene expression (PSGE), occurs in higher plants. Although the theory that PSGE solves genetic conflict between mother and father is widely accepted, De Jong and Scott (2007) argued that the conditions for PSGE to evolve are restricted. With respect to seed size, PSGE can only evolve when the developing offspring has a strong effect on its own resource acquisition.

Traditionally, RNA molecules are thought to function within the same cell in which they are generated. Recent evidence, however, reveal that, in addition to this conventional role, RNA also acts as a non-cell-autonomous signal molecule that travel intercellularly to regulate gene expression in remote target tissues. Kim, Canio, Kessler, and Sinha (2001) demonstrated that endogenous mRNA molecules not only travel between plant cells, but also is capable of causing phenotypic change in developing tissue. More interestingly, mRNAs have been reported not only in mammalian spermatozoa, but also in the male gametes of plants. For example Engel, Chaboud, Dumas, and MCormick (2003), showed that the sperm cells of

Zea mays have a complex complement of mRNAs. Recently, small RNAs have also been found in pollen (He, Yang, Wu, & Zheng, 2015; Martinez, 2017). Moreover, there is new evidence for small RNA-mediated communication between non-germ and germ cells in plants (Martinez, Panda, Koheler, & Slotkin, 2016). It has been suggested Darwin's so-called gemmules could include RNAs, particularly mRNA and small RNAs (Liu, 2006; Liu, 2008a). The new findings led me to assume that, during fertilization, pollen release mRNAs and small RNAs, which diffuse out into the tissues of the mother plant and there the translocated mRNAs and small RNAs cause changes in size, shape, colour, developmental timing and chemical composition of seeds and fruits, varying according to the particular male parent. This idea is consistent with Darwin's pangenetic explanation (Liu, 2008b).

2. TELEGONY: INFLUENCE OF PREVIOUS SIRE ON OFFSPRING

2.1 Darwin on Telegony

Telegony is the belief that the sire first mated to a female will have an influence upon some of that female's later offspring by another male. It is not only one of the most obscure problems in genetics, but also interests all sorts and conditions of men in many different parts of the world. Although the term "telegony" was coined by Weismann (1912), Darwin was the first to summarize this phenomenon in biological science. In Chapter 11 of his book *The Variation of Animals and Plants under Domestication*, Darwin collected many alleged examples of what he referred to as "the direct action of the male element on the female form". The classic case reported by Lord Morton (1821), was thus summarized by Darwin (1875) as follows:

> *A nearly purely-bred Arabian chestnut mare bore a hybrid to a quagga; she was subsequently sent to Sir Gore Ouseley, and produced two colts by a black Arabian horse. These colts were partially dun-coloured, and were striped on the legs more plainly than the real hybrid, or even than the quagga. One of the two colts had its neck and some other parts of its body plainly marked with stripes. Stripes on the body, not to mention those on the legs, are extremely rare,—I speak after having long attended to the subject,—with horses of all kinds in Europe, and are almost unknown in the case of Arabians. But what makes the case still more striking is that in these colts the hair of the mane resembled that of the quagga, being short, stiff, and upright. Hence there can be no doubt that the quagga affected the character of the offspring subsequently begot by the black Arabian horse.*

Another case cited by Darwin was that of a sow of Lord Western's black and white Essex strain, which Giles (1821) put first to a deep chestnut wild boar, and the offspring showed the characters of both parents, but in some the chestnut colour of the boar prevailed. After the chestnut wild boar had long been dead, the sow was put to a boar of her own black and white breed, a kind which was well known to breed very true and never to show any chestnut colour. Surprisingly, from the second union the sow produced some young plainly marked with the chestnut tint as in the first litter, which was never shown by the Essex strain. Darwin (1875) also gave cases of a hairless Turkish bitch which was mated by a spaniel, and had some hairless pups and some with short hair. She was subsequently mated with a hairless Turkish dog, but the offspring were as before. In addition Darwin (1875) gave many similar and well-authenticated facts, and concluded that "similar cases have so frequently occurred, that careful breeders avoid putting a choice female of any animal to an inferior male, on account of the injury to her subsequent progeny which may be expected to follow."

Darwin considered the phenomenon of telegony to be of special importance for understanding the mechanisms of heredity and development, and advanced this case as one of several in favour of his Pangenesis. He held that "it is certain that ovaria are sometimes affected by a previous impregnation, so that the ovules subsequently fertilized by a distinct male are plainly influenced in character". He thought that in case of Morton's mare it was "the diffusion, retention, and action of the gemmules included within the spermatozoa of the [first] male" that left the mare with a "quagga taint" that continued to manifest itself in her later offspring (Burkhardt, 1979; Darwin, 1875).

2.2 Further Evidence for Telegony

Although Galton rejected Darwin's Pangenesis, he treated telegony as a "well-ascertained fact" and explained it by his own theory of heredity (Burkhardt, 1979). After extensive investigations on this subject Romanes (1893, pp. 191–209) concluded that "my evidence is enough to prove the fact of a previous sire asserting his influence on a subsequent progeny, although this fact is one of comparatively rare occurrence." Now it appears that Romanes was right. In the scientific literature, there is indeed evidence for telegony, although the cases are much less than that of xenia (see Table 2).

In 1980s and 1990s, the phenomena of telegony were observed in the immune system of animals. According to Gorczynski et al. (1983), when

Table 2 Cases of Telegony in Animals and Human

Species	Characters Affected	References
Horse	Stripes	Morton (1821); Finn (1893)
Pig	Coat colour	Giles (1821); Lowe, 1896
	Body size	Hu (2005)
Dog	Hair	Darwin (1868)
	Wall-eyed	Cornevin (1891)
Cat	Tailless character	Lowe (1896)
Sheep	Flaps of skin	Darwin (1868); Finn (1893)
Mice	Immunity	Gorczynski et al. (1983)
Rabbit	Hair	Zhelnin (1964)
	Immunity	Sobey and Conolly (1986); Parer et al. (1995); Williams and Moore (1991)
Fly	Body size	Crean et al. (2014)
Beetle	Fecundity	Pascoal et al. (2018)
Human	Hypospadias	Cornevin (1891); Lingard (1884)
	Skin colour	Flint (1888)
	Deaf-mute	Cornevin (1891)
	Hair	Webber (1896); Lowe (1896)

normal females that had borne at least three litters to males made tolerant neonatally were subsequently mated to normal, non-immune males, their offspring demonstrated a hyporesponsive phenotype that did not differ from that of the progeny of fathers made tolerant neonatally. The response of this offspring was significantly lower than the response of mice born to normal females mated only to normal mates. Later Sobey and Conolly (1986) showed that male rabbits, mating after recovering from myxomatosis, could transmit immunity to offspring born to the female in the next seven months, including offspring sired by other males lacking immunity. This result was confirmed by later researchers (Parer, Sobey, Conolly, & Morton, 1995; Williams & Moore, 1991).

In recent years, there is evidence of telegony in insects. For example Crean, Kopps, and Bonduriansky (2014) demonstrated that adult body size of offspring could be influenced by the female's previous mate rather than the genetic sire in the neriid fly, *Telostylinus angusticollis*. In their previous experiments, it was found that males reared on a nutrient-rich larval diet (high-condition fathers) produce larger offspring than males reared on a nutrient-poor larval diet (low-condition fathers), indicating that males could transmit their environmentally acquired condition by paternal effects on body size of their offspring (Adler & Bonduriansky, 2013). Like many insects, young female neriid flies usually produce immature eggs that take several weeks to fully develop. Although males that mate with females during this interval sire have no offspring, it was hypothesized that they could still influence her future offspring. To test this telegony idea Crean et al. (2014) first mated immature female flies with two types of males: the first were well fed and in prime condition while the second were food deprived. Neither treatment group was expected to sire any offspring at all because the eggs were immature. Two weeks later, when their ovules were mature, they were mated for a second time with males of the two treatment groups and allowed to lay their fertilized eggs. Interesting, it was found that the size of the offspring of these twice-mated females was better predicted by the size of their first mate than by the size of the actual genetic father. If the first mate was big, so too were the offspring, even if the second mate was a runt, thus suggesting that it is possible for a male to transmit the character to his mate's subsequent offspring sired by another male. Most recently, a similar result was reported by Pascoal, Jarrett, Evans, and Kilner (2018).

2.3 Current Mechanistic Understanding of Telegony

2.3.1 Penetration of Sperm Into Somatic Cells and DNA Transformation

In the early 1910s, Kohlbrugge (1910, 1913) identified numerous spermatozoa in the mucosal cells of the uterus and uterine tube in the mouse, rabbit and bat. He held that modification of the mucosal cells could mediate a paternal influence upon the embryo, persisting from one pregnancy to the next and furnishing a basis for telegony. Decades later, Austin and colleagues examined sections of uteri of the Great Horseshoe bat and the Common Pipistrelle, and their observations were compatible with the possibility that the mucosa of the uterine-tube isthmus might be altered in some way by invading spermatozoa. They suggested that this could be brought about by genetic transformation (Austin & Bishop, 1959; Austin, 1959, 1960). Bendich, Borenfreund, and Sternberg (1974) reported that sperm could also penetrate cells in culture, raising the possibility that sperm could deliver genetic information into cells other than eggs. They demonstrated that mouse spermatozoa could penetrate into Chinese hamster fibroblast somatic cells, leading to a subsequent transfer of DNA from the sperm to the host cell nucleus. Brodsky and Ivanov (2009) suggested that spermatozoids might penetrate somatic cells *in vivo*, producing viable chimeric cells, which might survive in the body for a long time. All of these recent findings supported Darwin's Pangenetic explanation for telegony that the gemmules derived from the spermatozoa actually penetrate the nascent cells of the mother animal.

Holmgren et al. (1999) raised the question whether DNA could be transferred from one cell to another via the phagocytosis of apoptotic bodies. They showed that genomic DNA from apoptotic bodies was transferred to the nuclear compartment of phagocytosing cells and that this transferred DNA was stable over time. Thus apoptotic bodies derived from spermatozoa might be taken up by phagocytosing cells of the maternal tissues. It is reasonable to imagine that during coition, millions of DNA-containing spermatozoa are deposited in the body of the female and those not used in fertilization are absorbed by it. If this foreign DNA derived from spermatozoa could be integrated into somatic cells and influence ova, then later offspring might exhibit this influence in their genetic constitution, thus furnishing a basis for telegony (Liu, 2011).

2.3.2 The Effect of Telegony Mediated by Fetal Genes in Mother's Blood

The discovery of fetal genes in maternal plasma by Lo et al. (1997) not only makes it possible to determine the whole genome sequence of a fetus by using blood samples from the pregnant mother (Fan et al., 2012), but also opens the possibility that the young from a given mating are affected in some way by previous matings. During pregnancy, cells of the placenta shed DNA into the maternal bloodstream, so that the blood of the pregnant mother contains a mixture of her own genes and that of her fetus. It has been demonstrated that fetal DNA constitutes about 10% of the total DNA in maternal blood, and the minor allele fraction at loci with paternal-specific alleles is theoretically half of the fetal DNA fraction (Tsui & Lo, 2012). Since humans are diploid, meaning they have two copies of each chromosome, thus there are three haploid genomes in the mother's blood: her own two (one that she transmits to the fetus and the other that she does not) and the haploid genome derived from the father of the fetus (Bianchi, 2012). Fetal DNA may persist in mother's blood for many years, and even decades after pregnancy (Invernizzi et al. 2002). When and if fetal DNA derived from the father of the fetus is integrated into the genome of the mother's subsequent fetus by another male, it will probably affect the characters of the subsequent offspring. It is also possible that during pregnancy, whilst the fetus is developing and continuously shedding fetal DNA into mother's blood, the ovaries, and hence the primary oocytes, will stand a good chance of receiving the fetal DNA derived from the former father and so being transfected or transformed. When and if one of the transfected oocytes is afterwards fertilized by sperm from a different male, the resulting offspring is supposed to resemble the preceding male in some extent (Liu, 2013).

A question may arise to how can fetal DNA naturally transfect into an oocyte. Now direct microinjection of DNA has become a widely used technique for gene transfer. Gordon et al. (1980) showed that when DNA was microinjected into mouse oocytes, the injected DNA could be detected in a fraction of the fully developed mice. Further experiments showed that this DNA had been incorporated into the genome, and about 20% of mouse offspring routinely contain new transgenes (Gordon & Ruddle, 1981). Carballada, Degefa, and Esponda (2000) also used plasmid DNA in combination with cationic liposomes to transfect mouse eggs and embryos, and showed that immature eggs, collected from the ovaries, were easily transfected. It has been shown that early embryonic DNA contains multiple transient DNA breaks that could play a role in the incorporation of foreign DNA (Kiessling & Markoulaki, 2000). Generally speaking, transfection efficiency is low under natural conditions. Only a small portion of the cells become competent and a fraction of those are successful in absorbing foreign DNA. This may provide an explanation for the rare occurrence of telegony.

2.3.3 A Role for RNAs?

Over the past several decades, it has been documented that sperm nucleus contains a complex population of RNAs. A specific set of functional RNAs are delivered into oocytes at fertilization and are thought to contribute to early embryonic development (Godia, Swanson, & Krawetz, 2018; Hamatani, 2012; Ostermeier, Miller, Huntriss, Diamond, & Krawetz, 2004). It has also been shown that fetal RNA is present in the plasma of pregnant women. Vlkova, Szemes, Minarik, Turna, and Celec (2010). hypothesized that circulating fetal RNA in maternal blood might transfect maternal cells. Since DNA breaks usually precede genome rearrangement, RNA-templated DNA repair provides a direct mechanism for RNA-mediated genome rearrangement (Fang & Landweber, 2013). Thus the question arises whether the telegonic effect could be also caused by circulating fetal RNAs or spermatozoa RNAs. It has been proposed that spermatozoa may penetrate somatic tissues and could in theory deliver RNAs to somatic and germ cells where it could have regulatory and genetic effects (Liu, 2013). Recently, it has also been suggested that one of the possible functions of RNA in human seminal plasma might be to affect human oocytes as well as the offspring (Godia et al., 2018; Zagoskin, Davis, & Mukha, 2017), thus furnishing an alternative molecular basis for telegony.

Nearly 40 years ago, Steele presented a modern molecular view of Darwin's Pangenesis in his "somatic selection" hypothesis, which states that somatic information could be transmitted to the germline in the form of RNA and be transcribed to DNA by reverse transcriptase and become integrated into the germline DNA (Steele, 1979; Steele, Lindley, & Blanden, 1998). Spadafora (2008, 2018) described the phenomenon of sperm-mediated 'reverse' gene transfer, and believed that such a mechanism would lead to the generation and non-Mendelian propagation of new genetic information by mature spermatozoa. Thus retroviruses or retrotransposons may facilitate the reverse transcription of mRNA into cDNA leading to their integration into the female ovum genome and thus expression in the progeny (Liu, 2011).

3. BUD VARIATION

3.1 Darwin on Bud Variation

In the 11th chapter of *The Variation of Animals and Plants under Domestication*, Darwin (1868) examined several unusual forms of trait production, particularly bud variation, in which a plant produced a bud that grew into an offspring individual with highly distinctive characteristics. He also referred to this phenomenon as "sports", or "spontaneous variation" or "sudden variation" or "accidental variation" or "so-called spontaneous variation," and devoted considerable space to it (Johnson, 2015). Later de Vries referred to this phenomenon as mutation.

Darwin noticed that wild plants were so rarely liable to bud variation, while highly cultivated plants long propagated by artificial means had yielded many varieties by bud variation. He collected a good many cases of bud variation in different plants, such as peach, plum, cherry, grape, gooseberry, currant, pear, apple, banana, camellia, hawthorn, rose, dianthus, potato and so on. For example, he believed that nectarine was developed from peach by bud variation. He noticed that in potato, a single bud or eye sometimes varied and produced a new variety; or, occasionally, all the eyes in a tuber varied in the same manner and at the same time, so that the whole tuber assumed a new character. With regard to useful and ornamental trees Darwin (1875) wrote: "All the recorded varieties, as far as I can find out, have been suddenly produced by one single act of variation."

Concerning roses, he remarked on their marked tendency to "sport" and to produce varieties "not only by grafting and budding, but often by seed." He also described the segregation of the parental characters in seminal hybrids by bud variation, and called attention to the significant fact that many abrupt variations were not to be attributed either to reversion or to the splitting-up of hybrids. As he wrote:

> *Many cases of bud-variation, however, cannot be attributed to reversion, but to so-called spontaneous variability, as is so common with cultivated plants raised from seed. As a single variety of the Chrysanthemum has produced by buds six other varieties, and as one variety of the gooseberry has borne at the same time four distinct kinds of fruit, it is scarcely possible to believe that all these variations are due to reversion. We can hardly believe, as remarked in a previous chapter, that all the many peaches which have yielded nectarine-buds are of crossed parentage. Lastly, in such cases as that of the moss-rose, with its peculiar calyx, and of the rose which bears opposite leaves, in that of the Imatophyllum, etc., there is no known natural species or variety from which the characters in question could have been derived by a cross. We must attribute all such cases to the appearance of absolutely new characters in the buds. The varieties which have thus arisen cannot be distinguished by any external character from seedlings; this is notoriously the case with the varieties of the Rose, Azalea, and many other plants. It deserves notice that all the plants which have yielded bud-variations have likewise varied greatly by seed.*

In the summary of this chapter, Darwin wrote:

> *In the early half of this chapter I have given a long list of plants in which through bud-variation, that is, independently of reproduction by seed, the fruit has suddenly become modified in size, colour, flavour, hairiness, shape, and time of maturity; flowers have similarly changed in shape, colour, and doubleness, and greatly in the character of the calyx; young branches or shoots have changed in colour, in bearing spines, and in habit of growth, as in climbing and weeping; leaves have changed in colour, variegation, shape, period of unfolding, and in their arrangement on the axis. Buds of all kinds, whether produced on ordinary branches or on subterranean stems, whether simple or, as in tubers and bulbs, much modified and supplied with a stock of nutriment, are all liable to sudden variations of the same general nature.*

In a letter to Galton, Darwin wrote: "If gemmules were often deficient in buds, I cannot but think that bud-variations would be commoner than they are in a state of nature; nor does it seem that bud-variations often exhibit deficiencies which might be accounted for by the absence of the proper gemmules. I take a very different view of the meaning or cause of sexuality" (Darwin & Seward, 1903, p361). According to de Vries (1910), "Darwin believed the main branches of his genealogical tree to have arisen by a modification of his gemmules and that he regarded fluctuating variability as a phenomenon of an entirely different kind."

3.2 Further Evidence for Bud Variation

In our modern text book of fruit tree breeding, bud variation was introduced as a single chapter (Shen, 1980). Many of the present day varieties of horticultural plants owe their origin to bud variation, which can readily be propagated by asexual methods (Shamel & Pomeroy, 1936; Whitham & Slobodchikoff, 1981). Shamel and Pomeroy (1932) listed 1664 bud sports for a number of morphological or physiological characters in Citrus fruits and 234 bud sports in different fruit tree species. They noted that at that time 32% of the plant patents issued by the U. S. Patent Office were for plants originating as bud variations. It was also estimated that of the 8800 varieties cultivated in Europe in 1899, 5000 originated as bud variation (Shamel & Pomeroy, 1936).

In apple, a very high level of variation existed among the many bud-sports of Red Delicious and Starking (Dayton, 1964). Most of the known bud variations have to do with the colour of the fruit, probably because of the ease with which colour variations may be recognized (Strachan, 1934). Over the past several decades, hundreds of different Delicious apples have been introduced by bud variation (Meheriuk, 1977). In Japan, more than two-thirds of peach varieties are considered to be produced by bud variation from "Hakuto" or its relatives (Yamamoto, Mochida, & Hayashi, 2003a; Yamamoto et al., 2003b).

In 1906, a bud variation was found from one branch of a grapefruit tree which showed pink flesh. This single branch has since served as the parent stock for all the pink-fleshed grapefruit cultivated in the United States (Hartmann & Kester, 1975). Bud variation was found to occur in all potato varieties, some of them were superior to the parent stock. For example, in 1968, there were 20,000 cultivated acres of Red Craigs Royal potato, a variety by bud variation, being more than its parent stock of only 6000 cultivated acres (Howard, 1970).

In chapter 11 of *The Variation of Animal and Plants under Domestication*, when discussing the bud variation in dahlias, Darwin (1875) wrote: "Dahlias propagated by tubers under the hot climate of St. Domingo vary much; Sir R. Schomburgk gives the case of the 'Butterfly variety' which the second year produced on the same plant double and single flowers; here white petals edged with maroon; there of a uniform deep maroon. Mr. Bree also mentions a plant 'which bore two different kinds of self-coloured flowers, as well as a third kind which partook of both colors beautifully intermixed.' Another case is described of a dahlia with purple flowers which bore a white flower streaked with purple." Darwin's statement was confirmed by Shamel (1918), who made an inquiry amongst Southern California dahlia growers, and concluded that striking bud variations in dahlia plants were common and frequent occurrence. Sometimes two or more kinds of flowers appeared in the same plant. In many cases the interesting and valuable bud variations of the dahlia were isolated and propagated as new varieties. Thus many of the varieties were originated from bud variations.

3.3 The Cause of Bud Variation

As Imai (1935) pointed out, Darwin was the first to call general attention to the significance of bud variation. Based on Darwin's work, de Vries (1910) formulated his theory of mutation. One of the earlier cases of somatic mutation was reported by de Vries in variegated flowers of Antirrhinumii, which was later interpreted as due to a somatic mutation of genes (East, 1908; Emerson, 1922). The general understanding of the mechanism of bud variation was mainly based on de Vries's theory of mutation and Baur's theory of periclinal chimeras, which was supported and extended by later investigators. An alternative explanation of bud variation was the theory of somatic segregation, but it was not widely accepted (Imai, 1935).

Gene changes which occur in somatic cells can result in permanent changes in the plants in which they arise. These changes can be induced by point mutations, deletions, duplications, inversions, and so on (Hartmann & Kester, 1975). In some cases, plant parts arise from genetic changes in which all tissues have the same genetic makeup. If a somatic mutation arises in a meristematic cell of a bud, the derivative cells will also carry the mutation. Therefore a somatic mutation that occurs in the meristem of the bud can spread with the production. But in other cases, mutations often affect only a portion of the meristem. Plants or plant parts so derived are known as "chimeras" composed of two or more genetically distinct tissues. In this case a mutation must generally arise in a layer whose derivative cells are the floral parts of the plant for a somatic mutation to be expressed in the gametes of sexual reproduction (Neilson-Jones, 1969; Whitham & Slobodchikoff, 1981). In recent years, it has been shown that a high frequency of bud mutations in certain varieties may be caused by transposable elements (Inagaki, Hisatomi, Suzuki, Kasahara, & Iida, 1994; Uematsu et al., 2014).

4. REVERSION OR ATAVISM

Darwin (1868) discussed reversion or atavism in one long chapter entitled "Reversion or atavism" (Chapter 13) of *The Variation of Animals and Plants under Domestication.* He maintained that the tendency to reversion is often induced by a change of conditions and in the plainest manner by crossing, and regarded it as an essential part of the general law of inheritance. He conducted a series of experiments in animals, and the most striking example of reversion was with pigeon. He crossed a black Barb to a white Fantail, and also a black Barb to a red Spot, which was white pigeon with the tail and tail-coverts red, and a red spot on the forehead. He then crossed together the mongrel offspring obtained from these two crosses, and from this was produced a bird identical with *Columba livia* excepting that "the lead was tinted with a shade of red, evidently derived from the Spot, and was of a paler blue than in the rock-pigeon, as was the stomach." He also mated a White Silkie hen with a Spanish cock which was perfectly black except for the iridescent glossy black in hackle, saddle, wing bar and some of the tail feathers. "As the cocks grew old one ... became a gorgeous bird. When stalking about it closely resembled the wild Gallus bankiva, but with the red feathers rather darker." It is worth mentioning that Darwin's experiment was confirmed by Davenport (1910), who made the same cross with the same result.

Darwin's breeding experiments led him to think more seriously about reversion, and he knew that any theory of inheritance must explain how reversion could occur. According to his Pangenesis, a character could lie dormant only to be expressed after several generations. Thus reversion was explained as a consequence of long-dormant gemmules becoming active after the passage of many generations.

Belyaev, Ruvinsky, and Borodin (1981) reported a fused gene in mouse with about 6% of mutants produced wild type young, and after a few generations, 1% reverting to the mutant type. They proposed that inactive ("dormant") genes may be activated and reverted toward an inactive state. This explanation is similar to Darwin's Pangenesis. In addition Lolle, Victor, Young, and Pruitt (2005), showed that Arabidopsis plants homozygous for recessive mutant alleles of the organ fusion gene (HTH) could inherit allele-specific DNA sequence information that was not present in the chromosomal genome of their parents but was present in previous generations. They presented a model in which a type of stable RNA could be replicated and transmitted over multiple generations. This hypothesis is also consistent with Darwin's Pangenetic explanation (Liu, 2005).

5. INHERITANCE AND NON-INHERITANCE OF MUTILATION

In Chapter 12 of *The Variation of Animals and Plants under Domestication*, Darwin discussed the inheritance and non-inheritance of structures mutilated by injuries or altered by disease. He clearly stated that in most cases and normal conditions the effect of mutilation was not inherited:

> *Some mutilations have been practised for a vast number of generations without any inherited result. Godron remarks that different races of man have from time immemorial knocked out their upper incisors, cut off joints of their fingers, made holes of immense size through the lobes of their ears or through their nostrils, tatooed themselves, made deep gashes in various parts of their bodies, and there is no reason to suppose that these mutilations have ever been inherited. Adhesions due to inflammation and pits from the smallpox (and formerly many consecutive generations must have been thus pitted) are not inherited. With respect to Jews, I have been assured by three medical men of the Jewish faith that circumcision, which has been practised for so many ages, has produced no inherited effect.*
>
> ***Darwin (1875)***

Despite the above several negative cases, Darwin knew that there was also conclusive evidence that the effects of mutilation were sometimes inherited. He quoted the whole summary of Brown-Sequard's famous experiments showing that guinea-pigs of the next generation were affected by operations on certain nerves. He also gave many cases of inherited injuries:

> *With the lower animals Dr. Prosper Lucas has collected a long list of inherited injuries. A few instances will suffice. A cow lost a horn from an accident with consequent suppuration, and she produced three calves which were hornless on the same side of the head. With the horse, there seems hardly a doubt that exostoses on the legs, caused by too much travelling on hard roads, are inherited. Blumenbach records the case of a man who had his little finger on the right hand almost cut off, and which in consequence grew crooked, and his sons had the same finger on the same hand similarly crooked. A soldier, fifteen years before his marriage, lost his left eye from purulent ophthalmia, and his two sons were microphthalmic on the same side. … In all cases in which a parent has had an organ injured on one side, and two or more of the offspring are born with the same organ affected on the same side, the chances against mere coincidence are almost infinitely great. Even when only a single child is born having exactly the same part of the body affected as that of his injured parent, the chances against coincidence are great; and Professor Rolleston has given me two such cases which have fallen under his own observation,—namely of two men, one of whom had his knee and the other his cheek severely cut, and both had children born with exactly the same spot marked or scarred.*
>
> ***Darwin (1875)***

He recorded a case of inheritance of a peculiarity which arose from an injury accompanied by a diseased state of the part. "A gentleman, when a boy, had the skin of both thumbs badly cracked from exposure to cold, combined with some skin disease. His thumbs swelled greatly, and remained in this state for a long time. When they healed they were misshapen, and the nails ever afterwards were singularly narrow, short, and thick" (Darwin, 1881). According to him, this peculiarity was inherited to two of his four children, and two of his grandchildren. Darwin (1881) thought that it was reasonable to attribute the state of his thumbs to cold aided by skin disease, as he positively asserted that his thumbs were not originally misshapen, and there was no record of any previous inherited tendency of the kind in his family. Darwin (1873) also recorded a case about dogs with an inborn fear of butchers. Their father had a violent antipathy to butchers, probably as a result of being mistreated by someone, and this fear was transmitted by his offspring. From so many instances Darwin (1875) concluded that:

> *Finally, it must be admitted, more especially since the publication of Brown-Sequard's observations, that the effects of injuries, especially when followed by disease, or perhaps exclusively when thus followed, are occasionally inherited.*

Darwin's explanation of inherited mutilations, which as he notes, occur "especially or perhaps exclusively" when the injury has been followed by disease — is that all the representative gemmules, which would develop or repair or reproduce the injured part are attracted to the diseased surface during the reparative process and are there destroyed by the morbid action. The non-inheritance of mutilations was explained by persistence of gemmules from still earlier generations.

After Darwin, there was also evidence for the inheritance of mutilation in the scientific literature. For example, surgical or accident-induced side-by-side mirror-image doublets were experimentally produced in ciliates by a Chinese research group, and had been inherited stably for hundreds of generations (Tchang & Pang, 1965, 1977; Tchang, Shi, & Pang, 1964). This finding has since been confirmed by four independent laboratories, and extended by investigators worldwide for a variety of ciliates (Frankel, 2008; Grimes, McKenna, Goldsmith-Spoegler, & Knaupp, 1980; Yano & Suhama, 1991). The mirror image arrangement could be reliably generated by certain microsurgical operations (Frankel, 2008; Tchang et al., 1964). Davenport (1933) described a family showing two members with shortened metacarpals. Mrs. Ernestine R. Fernald, a young married woman, when about twelve years of age, sustained an injury to her hand by being hit with a croquet mallet, leading to the disappearance of the knuckle, a marked shortening of the finger, and a slight curvature of the finger. Interestingly, her daughter, Elizabeth Fernald, about ten years of age, showed an almost identical defect. This has been considered to be a case of the inheritance of mutilation. In a recent issue of *Science*, Laland (2018) mentioned the amoeba experiment described by Jennings (1937), who knocked out some teeth and found that the manipulated cells produced daughter cells with a similarly changed number of teeth, and alterations in tooth shape and size. As Laland (2018) commented, if Weismann had removed teeth from the amoeba *Difflugia corona*, he would have demonstrated reliable inheritance of mutilation.

Among the five anomalous facts discussed above, the existence of revision and bud variation had long been recognized by all geneticists. But three other phenomena, including xenia, telegony and the inheritance of mutilation, were doubted by such authorities as Weismann and Morgan. Over the past several decades, there has been increasing evidence for xenia, which is of great interest and importance in physiological research and plant production. The discoveries of cell-free fetal DNA, mobile RNAs and horizontal gene transfer also indicate that molecular mechanisms exist for telegony, one of the most controversial issues. Darwin's Pangenesis is the only theory that explains all the different types of phenomena.

REFERENCES

Adler, M. I., & Bonduriansky, R. (2013). Paternal effects on offspring fitness reflect father's social environment. *Evolutionary Biology, 40*, 288–292.

Austin, C. R. (1959). Entry of spermatozoa into the fallopian-tube mucosa. *Nature, 183*, 908–909.

Austin, C. R. (1960). Fate of spermatozoa in the female genital track. *Journal of Reproduction and Fertility, 1*, 151–156.

Austin, C. R., & Bishop, M. W. H. (1959). Presence of spermatozoa in the uterine-tube mucosa of bats. *Journal of Endocrinology, 18*, R7–R8.

Belyaev, D. K., Ruvinsky, A. O., & Borodin, P. M. (1981). Inheritance of alternative states of the fused gene in mice. *Journal of Heredity, 72*, 107–112.

Bendich, A., Borenfreund, E., & Sternberg, S. S. (1974). Penetration of somatic mammalian cells by sperm. *Science, 183*, 857–859.

Bianchi, D. W. (2012). Prenatal diagnostics: Fetal genes in mother's blood. *Nature, 487*, 304–305.

Brodsky, S. V., & Ivanov, I. (2009). Spermatozoa-somatic cell fusion – a mechanism for micro-chimerism formation. *Journal of Theoretical Biology, 259*, 190–192.

Bulant, C., & Gallais, A. (1998). Xenia effects in maize with normal endosperm: I. Importance and stability. *Crop Science, 38*, 1517–1525.

Burkhardt, R. W. (1979). Closing the door on Lord Morton's mare: The rise and fall of telegony. *Studies in History of Biology, 3*, 1–21.

Carballada, R., Degefa, T., & Esponda, P. (2000). Transfection of mouse eggs and embryos using DNA combined to cationic liposomes. *Molecular Reproduction and Development, 56*, 360–365.

Cornevin, C. H. (1891). *Traite de Zootechnie Generale*. Paris: Bailliere.

Crean, A. J., Kopps, A. M., & Bonduriansky, R. (2014). Revisiting telegony: Offspring inherit an acquired characteristics of their mother's previous mate. *Ecology Letters, 17*, 1545–1552.
Dag, A., & Mizrahi, Y. (2005). Effect of pollination method on fruit set and fruit characteristics in the vine cactus Selenicereus megalanthus ("yellow pitaya"). *The Journal of Horticultural Science and Biotechnology, 80*, 618–622.
Darwin, C. (1868). *The variation of animals and plants under domestication*. London: John Murray.
Darwin, C. (1873). Inherited instinct. *Nature, 7*, 281.
Darwin, C. (1875). *The variation of animals and plants under domestication* (2nd ed.). London: John Murray.
Darwin, C. (1881). Inheritance. *Nature, 24*, 257.
Darwin, F. (1888). *The life and letters of Charles Darwin*. London: John Murray.
Darwin, F., & Seward, A. C. (Eds.). (1903). *More letters of Charles Darwin* (p. 361). London: John Murray.
Davenport, C. B. (1910). The new views about reversion. *Proceedings of the American Philosophical Society, 49*, 291–296.
Davenport, C. B. (1933). An alleged case of inheritance of acquired characters. *The American Naturalist, 67*, 549–558.
Dayton, D. F. (1964). Variation in the pattern of red color distribution in the skin of delicious and Starking bud sports varieties. *Proceedings of the American Society for Horticultural Science, 84*, 44–50.
De Jong, T. J., & Scott, R. J. (2007). Parental conflict does not necessarily lead to the evolution of imprinting. *Trends in Plant Science, 12*, 439–443.
de Vries H. (1910). *The mutation theory: Experiments and observations on the origin of species in the vegetable kingdom, The origin of species by mutation Vol. I.* English translation by J. B. Farmer & A. D. Darbishire (pp. 38–39). London: Kegan Paul, Trench, Trubner & Co. Ltd.
Denney, J. O. (1992). Xenia includes metaxenia. *HortScience, 27*, 722–727.
Duan, M.-X., Song, T.-M., Wang, L.-M., Fan, H.-W., & Zhao, J.-R. (2002). Study on the xenia effect of high oil corn. *Acta Agronomica Sinica, 28*, 208–214.
Duc, G., Moessner, A., Moussy, F., & Mousset-Declas, C. (2001). A xenia effect on number and volume of cotyledon cells and on seed weight in faba bean (*Vicia faba* L.). *Euphytica, 117*, 169–174.
Dunn, L. C. (1973). Xenia and the origin of genetics. *Proceedings of the American Philosophical Society, 117*, 105–111.
East, E. M. (1908). Suggestions concerning certain bud variations. *The Plant World, 11*, 77–83.
Emerson, R. A. (1922). The nature of bud variations as indicated by their mode of inheritance. *The American Naturalist, 41*, 64–79.
Engel, M. L., Chaboud, A., Dumas, C., & MCormick, S. (2003). Sperm cells of Zea mays have a complex complement of mRNAs. *The Plant Journal, 34*, 697–707.
Fang, W., & Landweber, L. F. (2013). RNA-mediated genome rearrangement: Hypothesis and evidence. *BioEssays: News and Reviews in Molecular, Cellular and Developmental Biology, 35*, 84–87.
Fan, H.-C., Gu, W., Wang, J., Blumenfeld, Y. J., El-Sayed, Y. Y., & Quake, S. R. (2012). Non-invasive prenatal measurement of the fetal genome. *Nature, 487*, 320–324.
Fattahi, R., Mohammadzedeh, M., & Khadivi-Khub, A. (2014). Influence of different pollen sources on nut and kernel characteristics of hazelnut. *Scientia Horticulturae, 173*, 15–19.
Finn, F. (1893). Some facts of telegony. *Natural Science, 3*, 436–446.
Flint, A. (1888). *A text-book of human physiology* (p. 797). New York: D. Appleton & Co.
Focke, W. O. (1881). *Die Pflanzen-Mischlinge: ein Beitrag zur Biologie der Gewächse*. Berlin: Bomtraeger.

Frankel, J. (2008). What do genic mutations tell us about the structural patterning of a complex single-celled organism? *Eukaryotic Cell, 7*, 1617–1639.
Freytag, G. F. (1979). Metaxenia effects on pod size development in the common bean. *Journal of Heredity, 70*, 444–446.
Gharaqhani, A., Soloklui, A. A. G., Oraquzie, N., & Zare, D. (2017). Pollen influences fruit quality, aril properties, and seed characteristics in pomegranate. *International Journal of Fruit Science, 17*, 333–348.
Giles, D. (1821). Particulars of fact, nearly similar to that related by Lord Morton. *Philosophical Transactions of the Royal Society, 111*, 23–24.
Godia, M., Swanson, G., & Krawetz, S. A. (2018). A history of why father's RNA matters. *Biology of Reproduction*. https://doi.org/10.1093/biolre/ioy007 (in press).
Gorczynski, R. M., Kennedy, M., Macrae, S., & Ciampi, A. (1983). A possible maternal effect in the abnormal hyporesponsiveness to specific alloantigens in offspring born to neonatally tolerant fathers. *The Journal of Immunology, 131*, 1115–1120.
Gordon, J. W., & Ruddle, F. H. (1981). Integration and stable germ line transmission of genes injected into mouse pronuclei. *Science, 214*, 1244–1246.
Gordon, J. W., Scangos, G. A., Plotkin, D. J., Barbosa, J. A., & Ruddle, F. H. (1980). Genetic transformation of mouse embryos by microinjection of purified DNA. *Proceedings of the National Academy of Sciences of the United States of America, 77*, 7380–7384.
Grimes, G. W., McKenna, M. E., Goldsmith-Spoegler, C. M., & Knaupp, E. A. (1980). Patterning and assembly of ciliature are independent processes in hypotrich ciliates. *Science, 209*, 281–283.
Gupta, A., Godara, R. K., Panda, A. K., Sharma, S., & Kaushik, L. (2017). Effect of pollen sources on quality characteristics of different cultivars of date palm (*P. dactylifera* L.) under Haryana Conditions. *International Journal of Current Microbiology and Applied Sciences, 6*, 71–77.
Hamatani, T. (2012). Human spermatooal RNAs. *Fertility and Sterility, 97*, 275–281.
Hartmann, H. T., & Kester, D. E. (1975). *Plant propagation: Principles and practices*. Englewood Cliffs, New Jersey: Prentice-Hall.
He, H., Yang, T. Y., Wu, W. Y., & Zheng, B. L. (2015). Small RNAs in pollen. *Science China. Life Sciences, 58*, 246–252.
Holmgren, L., Szeles, A., Rajnavolgyi, E., Folkman, J., Klein, G., Ernberg, I., & Falk, K. I. (1999). Horizontal transfer of DNA by the uptake of apoptotic bodies. *Blood, 93*, 3956–3963.
Howard, H. W. (1970). *Genetics of the potato*. Berlin Heidelberg New York: Springer.
Hu, J.-Y. (2005). *Swine Genetics and Breeding* (pp. 147–150). Beijing: China Agricultural University Press.
Imai, Y. (1935). The mechanism of bud variation. *The American Naturalist, 69*, 587–595.
Inagaki, Y., Hisatomi, Y., Suzuki, T., Kasahara, K., & Iida, S. (1994). Isolation of a Suppressor-mutator/Enhancer-like transposable element, Tpnl, from Japanese morning glory bearing variegated flowers. *The Plant Cell, 6*, 375–383.
Invernizzi, P., Biondi, M. L., Battezzati, P. M., Perego, F., Selmi, C., Cecchini, F., … Simoni, G. (2002). Presence of fetal DNA in maternal plasma decades after pregnancy. *Human Genetics, 110*, 587–591.
Jennings, H. S. (1937). Formation, inheritance and variation of the teeth in *Difflugia corona*. A study of the morphogenic activities of rhizopod protoplasm. *Journal of experimental Zoology, 77*, 287–336.
Johnson, C. (2015). *Darwin's Dice: The Idea of Chance in the Thought of Charles Darwin*. Oxford and New York: Oxford University Press.
Kermicle, J. L. (1978). Imprinting of gene action on maize endosperm. In D. B. Walden (Ed.), *Maize breeding and genetics* (pp. 357–371). New York: Wiley.
Kiesselbach, T. A., & Leonard, W. H. (1932). The effect of pollen source upon the grain yield of corn. *Journal of American Society of Agronomy, 24*, 517–523.

Kiessling, A. A., & Markoulaki, S. (2000). Integration of gametes with exogenous genes: Possible opportunities for incorporation into embryonic genome. *Molecular Reproduction and Development, 56*, 271–274.
Kim, M., Canio, W., Kessler, S., & Sinha, N. (2001). Developmental changes due to long-distance movement of a homebox fusion transcript in tomato. *Science, 293*, 287–289.
Kodad, O., Estopanan, G., Juan, T., & Company, R. S. I. (2009). Xenia effect on oil content and fatty acid and tocopherol concentrations in autogamous almond cultivars. *Journal of Agricultural and Food Chemistry, 57*, 10809–10813.
Kohlbrugge, J. H. F. (1910). Der Einfluss der Spermatozoiden auf den Uterus. Ein Beitrag zur Telegonie. *Zeitschrift für Morphologie und Anthropologie, 12*, 359–368.
Kohlbrugge, J. H. F. (1913). Die Verbreitung der Spermatozoden im weiblichen Korper und m befruchteten Ei. *Archiv für Entwicklungsmechanik der Organismen, 35*, 165–188.
Laland, K. N. (2018). Beyond epigenetics. *Science, 360*, 1408.
Lingard, A. (1884). Hereditary transmission of hypospadias and its transmission by indirect atavism. *Lancet, 123*, 703.
Liu, Y.-S. (2005). Reversion: Going back to Darwin's works. *Trends Plant Sciences, 10*, 459–460.
Liu, Y.-S. (2006). Response to till Bottraud and Gaggiotti: Going back to Darwin's work. *Trends in Plant Science, 11*, 472–473.
Liu, Y.-S. (2008a). A new perspective on Darwin's Pangenesis. *Biological Reviews of the Cambridge Philosophical Society, 83*, 141–149.
Liu, Y.-S. (2008b). A novel mechanism for xenia? *HortScience, 43*, 706.
Liu, Y.-S. (2011). Telegony, the sire effect and non-mendelian inheritance mediated by spermatozoa: A historical overview and modern mechanistic speculations. *Reproduction in Domestic Animals, 46*, 338–343.
Liu, Y. S. (2013). Fetal genes in mother's blood: A novel mechanism for telegony? *Gene, 524*, 414–416.
Lo, Y. M. D., Corbetta, N., Chamberlain, P. F., Rai, V., Sargent, I. L., Redman, C. W. G., & Wainscoat, J. S. (1997). Presence of fetal DNA in maternal plasma and serum. *The Lancet, 350*, 485–487.
Lolle, S. J., Victor, J. L., Young, J. M., & Pruitt, R. E. (2005). Genome-wide non-Mendelian of extra-genomic in formation in. *Arabidopsis. Nature, 434*, 505–509.
Lowe, E. J. (1896). Telegony. *The British Medical Journal, 2*, 1352.
Lukic, M., Maric, S., Milosevic, N., & Mitrovic, O. (2015). Effect of metaxenia on pomological traits of "Topaz" apple cultivar. *Acta Horticulturae, 1139*, 329–334.
Martinez, G. (2017). Isolation and detection of small RNAs from pollen. *Methods in Molecular Biology, 1669*, 237–250.
Martinez, G., Panda, K., Koheler, C., & Slotkin, R. K. (2016). Silencing in sperm cells is directed by RNA movement from the surrounding nurse cell. *Nature Plants, 2*, 16030.
McKay, J. W., & Crane, H. L. (1939). Xenia in the chestnut. *Science, 89*, 348–349.
Meheriuk, M. (1977). Maturity studies on red sports of delicious apples. *Canadian Journal of Plant Science, 57*, 963–967.
Mizrahi, Y., Mouyal, J., Nerd, A., & Sitrit, Y. (2004). Metaxenia in the vine cacti *Hylocereus polyrhizus* and *Selenicereus* spp. *Annals of Botany, 93*, 469–472.
Mohammadi, N., Rastgoo, S., & Izadi, M. (2017). The strong effect of pollen source and pollination time on fruit set and the yield of tissue culture-derived date palm (*Phoenix dactylifera* L.) trees cv. Barhee. *Scientia Horticulturae, 224*, 343–350.
Morton, L. (1821). A communication of a singular fact in natural history. *Philosophical Transactions of the Royal Society, 111*, 20–22.
Nebel, B. R., & Trump, J. I. (1932). Xenia and metaxenia in apples. *Proceedings of the National Academy of Sciences of the United States of America, 18*, 356–359.
Neilson-Jones, W. (1969). *Plant chimeras* (2nd ed.). London: Methuen.

Nixon, R. W. (1928). The direct effect of pollen on the fruit on the date palm. *Journal of Agricultural Research, 36*, 97–128.
Nixon, R. W. (1934). Metaxenia in dates. *Proceedings of the Society for Horticultural Science, 32*, 221–226.
Ostermeier, G. C., Miller, D., Huntriss, J. D., Diamond, M. P., & Krawetz, S. A. (2004). Delivering spermatozoan RNA to the oocyte. *Nature, 429*, 154.
Pahlavani, M. H., & Abolhasani, K. (2006). Xenia effect on seed and embryo size in cotton. *Journal of Applied Genetics, 47*, 331–335.
Parer, I., Sobey, W. R., Conolly, D., & Morton, R. (1995). Sire transmission of acquired resistance to myxomatosis. *Australian Journal of Zoology, 43*, 459–465.
Pascoal, S., Jarrett, B. J. M., Evans, E., & Kilner, R. M. (2018). Superior stimulation of female fecundity by subordinate males provides a mechanism for telegony. *Evolution letters, 2*, 114–125.
Piotto, F. A., Bataqin-Piotto, K. D., de Almeida, M., & Oliveira, G. C. X. (2013). Interspecific xenia and metaxenia in seeds and fruits of tomato. *Scientia Agricola, 70*, 102–107.
Qiu, Y.-P., Dai, H.-F., Li, Z.-Q., Ou, L., Xiang, X., Chen, J.-Z., & Wang, B.-X. (2006). Effects of pollinator on fruit quality of Guiwei litchi cultivar. *Journal of Fruit Science, 23*, 703–706.
Romanes, G. J. (1893). *An examination of Weismannism*. London, Longmans: Green & Co.
Sabir, A. (2015). Xenia and metaxenia in grapes: Differences in berry and seed characteristics of maternal grape cv. 'Narince' (*Vitis vinifera* L.) as influenced by different pollen sources. *Plant Biology, 17*, 567–573.
Salomon-Torresl, R., Ortiz-Uribe, N., Villa-Angulo, R., Villa-Angulo, C., Norzagaray-Plasencia, S., & Garcia-Verdugo, C. D. (2017). Effect of pollenizers on production and fruit characteristics of date palm *(Phoenix dactylifera* L.*)* cultivar Medjool in Mexico. *Turkish Journal of Agriculture and Forestry, 41*, 338–347.
Sanchez-Perez, R., Arrazola, G., Martin, M. L., Grane, N., & Dicenta, F. (2012). Influence of the pollinizer in the amygdalin content of almonds. *Scientia Horticulturae, 139*, 62–65.
Seal, A. G., Dunn, J. K., de Silva, H. N., McGhie, T. K., & Lunken, R. C. M. (2013). Choice of pollen parent affects red flesh colour in seedlings of diploid *Actinidia chinensis* (kiwifruit). *New Zealand Journal of Crop and Horticultural Science, 41*, 207–218.
Shamel, A. D. (1918). Bud variation in dahlias. *The Journal of Heredity, 9*, 362–364.
Shamel, A. D., & Pomeroy, C. S. (1932). Bud variation in apples. *The Journal of Heredity, 23*, 173–180/213–221.
Shamel, A. D., & Pomeroy, C. S. (1936). Bud mutations in horticultural crops. *The Journal of Heredity, 27*, 486–494.
Shen, D.-X. (1980). *Fruit tree breeding*. Shanghai Science and Technology Press.
Sobey, W. R., & Conolly, D. (1986). Myxomatosis: Non-genetic aspects of resistance to myxomatosis in rabbits, *Oryctolagus cuniculus*. *Australian Wildlife Research, 13*, 177–187.
Spadafora, C. (2008). Sperm-mediated 'reverse' gene transfer: A role of reverse transcriptase in the generation of new genetic information. *Human Reproduction, 23*, 735–740.
Spadafora, C. (2018). The "evolutionary field" hypothesis. Non-Mendelian transgenerational inheritance mediates diversification and evolution. *Progress in Biophysics and Molecular Biology, 134*, 27–37.
Steele, E. J. (1979). *Somatic selection and adaptive Evolution: On the inheritance of acquired characters*. Toronto: Williams-Wallace.
Steele, E. J., Lindley, R. A., & Blanden, R. V. (1998). *Lamarck's Signature: How retrogenes are changing Darwin's natural selection paradigm*. Massachusetts: Perseus Books.
Strachan, C. C. (1934). Colour strains of the Delicious apple. *Scientific Agriculture, 14*, 384–399.
Swingle, W. T. (1928). Metaxenia in the date palm — possibly a hormone action by the embryo or endosperm. *Journal of Heredity, 19*, 257–268.

Tchang, T.-R., & Pang, Y.-B. (1965). Conditions for the artificial induction of monster jumelles of *Stylonychia mytilus* which are capable of reproduction. *Scientia Sinica, 14*, 1332–1338.

Tchang, T.-R., & Pang, Y.-B. (1977). The cytoplasmic differentiation of jumelle *Stylonychia*. *Scientia Sinica, 20*, 234–243.

Tchang, T.-R., Shi, X.-B., & Pang, Y.-B. (1964). An induced monster ciliate transmitted through three hundred and more generations. *Scientia Sinica, 13*, 850–853.

Tsui, N. B. Y., & Lo, Y. M. D. (2012). Recent advances in the analysis of fetal nucleic acids in maternal plasma. *Current Opinion in Hematology, 19*, 462–468.

Uematsu, C., Katayama, H., Makino, I., Inagaki, A., Arakawa, O., & Martin, C. (2014). Peace, a MYB like transcription factor, regulates petal pigmentation in flowering peach 'Genpei' bearing variegated and fully pigmented flowers. *Journal of Experimental Botany, 65*, 1081–1094.

Vlkova, B., Szemes, T., Minarik, G., Turna, J., & Celec, P. (2010). Circulating free fetal nucleic acids in maternal plasma and preeclampsia. *Medical Hypothesis, 74*, 1030–1032.

Wallace, H. M., & Lee, L. S. (1999). Pollen source, fruit set and xenia in mandarins. *The Journal of Horticultural Science and Biotechnology, 74*, 82–86.

Wallace, H. M., Vithanage, V., & Exlev, E. M. (1996). The effect of supplementary pollination on nut set of *Macadamia* (Proteaceae). *Annals of Botany, 78*, 765–773.

Webber, H. W. (1896). Telegony. *Bmj: British Medical Journal, 2*, 1083.

Weingartner, U., Camp, K.-H., & Stamp, P. (2004). Impact of male sterility and xenia on grain quality traits of maize. *European Journal of Agronomy, 21*, 239–247.

Weingartner, U., Kaeser, O., Long, M., & Stamp, G. (2002). Combing cytoplasmic male sterility and xenia increases grain yield of Maize hybrids. *Crop Science, 42*, 1848–1856.

Weismann, A. (1912). *The germ-plasm—a theory of heredity*. New York: The Walter Scott Publishing Co., LTD.

Whitham, T. G., & Slobodchikoff, C. N. (1981). Evolution by individuals, plant–herbivore interactions, and mosaics of genetic variability: The adaptive significance of somatic mutations in plants. *Oecologia, 49*, 287–292.

Williams, C. K., & Moore, R. J. (1991). Inheritance of acquired immunity to myxomatosis. *Australian Journal of Zoology, 39*, 307–311.

Xie, L.-H., Niu, L.-X., Zhang, Y.-L., Jin, M., Ji, D., & Zhang, X.-X. (2017). Pollen sources influences the traits of seed and seed oil in Paeonia ostii. *HortScience, 52*, 700–705.

Yamamoto, T., Mochida, K., & Hayashi, T. (2003a). Shanghai Suimitsuto, one of the origins of Japanese peach cultivars. *Journal of the Japanese Society for Horticultural Science, 72*, 116–121.

Yamamoto, T., Mochida, K., Imai, T., Haji, T., Yaegaki, H., Yamaguchi, M., ... Hayashi, T. (2003b). Parentage analysis in Japanese peaches using SSR markers. *Breeding Science, 53*, 35–40.

Yang, L.-F., Yao, L.-F., Zhou, X.-M., & Yang, L.-H. (2002). Xenia in the apricot. *Journal of Fruit Science, 19*, 275–277.

Yano, J., & Suhama, M. (1991). Pattern formation in mirror-image doublets of the hypotrich ciliate Stylonychia pustulata. *Journal of Protozoology, 38*, 111–121.

Zagoskin, M. V., Davis, R. E., & Mukha, D. V. (2017). Double stranded RNA in human seminal plasma. *Frontiers in Genetics, 8*, 154.

Zhelnin, V. A. (1964). The phenomenon of telegony in rabbits. *Krolikovodstvo Zvehovodstvo, 7*, 20–21.

Zurawicz, E., Studnicki, M., Kubik, J., & Pruski, K. (2018). A careful choice of compatible pollinizers significantly improves the size of fruits in red raspberry (*Rubus idaeus* L.). *Scientia Horticulturae, 235*, 253–257.

CHAPTER FIVE

Natural Selection and Pangenesis: The Darwinian Synthesis of Evolution and Genetics

Yongsheng Liu*,§
*Henan Collaborative Innovation Center of Modern Biological Breeding, Henan Institute of Science and Technology, Xinxiang, China
§Department of Biochemistry, University of Alberta, Edmonton, AB, Canada
E-mail: ysliu63@yahoo.ca.

Contents

Advances in Genetics, Volume 102
ISSN 0065-2660
https://doi.org/10.1016/bs.adgen.2018.05.010

Abstract

Since the end of the 19th century, Lamarck's name has been tightly linked to the notion of the inheritance of acquired characters. Darwin regarded Lamarck as a great zoologist and a forerunner of evolution, and repeatedly expressed the opinion that "natural selection has been the main but not the exclusive means of modification." The original Darwinism not only includes natural selection, but also the inheritance of acquired characters and mutation. Neo-Darwinism considers natural selection as the one controlling process of evolution, but denies the inheritance of acquired characters. Lysenkoism accepts the inheritance of acquired characters and graft hybridization, but denies the significance of Malthusism and Mutationism. It has been suggested that the "modern synthesis", which evolved from neo-Darwinism, needs a rethink. I propose that there is a need to go back to Darwin's own synthesis which combined his theory of evolution by natural selection with his theory of heredity and variation — Pangenesis.

1. LAMARCKISM AND THE INHERITANCE OF ACQUIRED CHARACTERS

1.1 Lamarck and Lamarckism

Jean Baptiste Lamarck (1744—1829) was one of the great scientists of his age (Stafleu, 1971) . His legacy has been well discussed in the literature (Gadjev, 2015; Jablonka & Lamb, 1995; Steele, Lindley, & Blanden, 1998). He was the first to study invertebrate animals systematically, and a founder of invertebrate zoology (Balter, 2000). He assimilated his broad experience into a cosmic system which incorporated a general theory of evolution, providing the first consistent explanation of the relations between, and the adaptations of, living organisms. It was Lamarck who coined the term "biology" (Stafleu, 1971). It is not, however, as the founder of biology that he is remembered today. Historians of biological science have pointed to the irony — that the term Lamarckism (Fine, 1979).

Lamarckism is considered as the proposition that organisms respond to their environment purposefully by the formation of modifications which are adaptive to that environment. It is based on four principles: (1) living organisms and their component parts tend to increase in size; (2) new organs result from new needs and from the movements which these needs cause; (3) use of an organ leads to its further development, and disuse leads to its degeneration; (4) modifications produced by the actions of these principles are transmitted to future generations, and their cumulative effects over long periods of time are responsible for the origin of species (Dodson, 1961).

1.2 Lamarck and the Inheritance of Acquired Characters

As Burkhardt (2013) pointed out, since the end of the 19th century, Lamarck's name has been tightly linked to the notion of the inheritance of acquired characters. In 1809, in his famous *Philosophie zoologique*, Lamarck set out this idea more systematically in the form of two laws:

> *First Law: In every animal that has not reached the end of its development, the more frequent and sustained use of any organ will strengthen this organ little by little, develop it, enlarge it, and give to it a power proportionate to the duration of its use; while the constant disuse of such an organ will insensibly weaken it, deteriorate it, progressively diminish its faculties, and finally cause it to disappear.*
>
> *Second Law: All that nature has caused individuals to gain or lose by the influence of the circumstances to which their race has been exposed for a long time, and, consequently, by the influence of a predominant use or constant disuse of an organ or part, is conserved through generation in the new individuals descending from them, provided that these acquired changes are common to the two sexes or to those which have produced these new individuals.*
>
> ***Lamarck (1809)***

In his recent article, Burkhardt (2013) made a number of comments regarding Lamarck and the inheritance of acquired characters: (1) Lamarck was not the first to present the idea of the inheritance of acquired characters, and he himself never claimed it as his own. (2) Lamarck maintained that he was the first to recognize "the importance of this law and the light it sheds on the causes that have led to the astonishing diversity of animals." (3) Lamarck believed that for characters acquired as the result of new habits to be inherited, they needed to have been acquired by both parents. (4) Lamarck believed that change was slow and incremental, and could be expected only if changed circumstances caused animals to adopt new habits. (5) The inheritance of acquired characters was only a small item in the broad scheme of Lamarck's theorizing. (6) Lamarck's thinking has long been characterized through his examples of the inheritance of acquired characters, such as his giraffe example. (7) Lamarck's examples and his model of species change were about functional, adaptive changes. (8). Lamarck offered no explanation of the mechanisms by which acquired characters were inherited. (9) Lamarck himself never used the phrase "the inheritance of acquired characters." (10) Darwin, like Lamarck, was a firm believer in the inheritance of acquired characters.

1.3 Lamarck's Failure to Explain the Inheritance of Acquired Characters

To my knowledge, the recent article by Burkhardt (2013) is comprehensive and objective, the following statement by him is particularly appreciated:

> *Lamarck, in contrast, as we have seen, had not offered any particular hypothesis about how acquired characters might be transmitted from one generation to the next. His focus was on the changes that individuals underwent as the result of adopting new habits. His emphasis, in other words, was on the use and disuse side of the equation (law 1 of his* Philosophie zoologique*), not the hereditary transmission side of it (law 2). He did not identify a range of different hereditary phenomena in need of explanation. ... It was Darwin, not Lamarck, who offered a theory to explain how characters acquired as the result of use and disuse were transmitted from one generation to the next. It was also Darwin, as we have seen, who allowed that he had provided more evidence on the effects of use and disuse than anyone before him.*

According to Lamarck, animals responded to new needs required by the novel circumstances of a changing environment by adaptive efforts. These efforts resulted in changing habits and functions, which were transmitted to the offspring. Lamarck assumed that the vectors of change within the organisms were the so-called "fluids." He held that the "subtle fluids," emitted by the sense organs and circulating through the nerves induced modifications of the body structures (Por, 2006). Today this may seem to be a nebulous and crude hypothesis. But we must understand that by the time Lamarck presented this hypothesis, the cell theory had not been developed. Actually, before the cell theory was presented, Lamarck (1809) stated clearly that all plants and animals were composed essentially of cellular tissue, without which "no living body would be able to exist nor could have been formed." But Lamarck's clear and positive statement of the fundamental importance of cellular tissue was not supported by facts (Gerould, 1922).

2. DARWINISM INCLUDES LAMARCKIAN INHERITANCE

2.1 Darwin's Comment on Lamarck

Darwin's comments were complex, and sometimes self-contradictory. He made numerous short marginal annotations in the first volume of Lamarck's *Zoological Philosophy*, as well as longer, more comprehensive notes at the ends of chapters. He claimed that it was a "Very poor & useless Book", and did not like Lamarck's non-approach to the issue of species (Glick & Kohn, 1996). In a letter to Lyell, he wrote that Lamarck's book "appeared to me extremely poor; I got not a single fact or idea from it." Darwin complained that the geologist referred "repeatedly to my view as a modification of Lamarck's doctrine of development and progression. If this is your deliberate opinion there is nothing to be said, but it does not seem so to me. Plato, Buffon, my grandfather before Lamarck, and others, propounded the obvious views that if species were not created separately, they must have descended from other species, and I can see nothing else in common between the *Origin* and Lamarck." (Glick & Kohn, 1996).

In his *The Origin of Species*, however, Darwin (1872) gave Lamarck sufficient credit, acknowledging him as a great zoologist and a forerunner of evolution:

> *Lamarck was the first man whose conclusions on the subject excited much attention. This justly-celebrated naturalist first published his views in 1801; he much enlarged them in 1809 in his "*Philosophie Zoologique,*" and subsequently, in 1815, in the Introduction to his "*Hist. Nat. des Animaux sans Vertébres.*" In these works he upholds the doctrine that species, including man, are descended from other species. He first did the eminent service of arousing attention to the probability of all change in the organic, as well as in the inorganic world, being the result of law, and not of miraculous interposition. Lamarck seems to have been chiefly led to his conclusion on the gradual change of species, by the difficulty of distinguishing species and varieties, by the almost perfect gradation of forms in certain groups, and by the analogy of domestic productions. With respect to the means of modification, he attributed something to the direct action of the physical conditions of life, something to the crossing of already existing forms, and much to use and disuse, that is, to the effects of habit. To this latter agency he seemed to attribute all the beautiful adaptations in nature; such as the long neck of the giraffe for browsing on the branches of trees. But he likewise believed in a law of progressive development; and as all the forms of life thus tend to progress, in order to account for the existence at the present day of very simple productions, he maintains that such forms were now spontaneously generated.*

Darwin held that Lamarckian inheritance was essential in accounting for many difficult problems in evolution, such as the inordinate growth of the human brain, the loss of structures through disuse, the origin of many instincts, and so on. He reaffirmed his Lamarckism in the preface to *The Descent of Men*, where he complained that his critics had overlooked this element in his work. There he reiterated that he had always maintained "that great weight must be attributed to the inherited effects of use and disuse, with respect both to the body and mind." (Cochrane, 1997; Darwin, 1874).

2.2 Darwinism Includes Natural Selection, Lamarckian Inheritance and Mutation

As Simpson (1960) correctly pointed out, the heart of Darwin's explanation of evolution was natural selection. Darwin always considered this as his most important contribution. However, from the first edition of *The Origin of Species* Darwin expressed the opinion that "natural selection has been the main but not the exclusive means of modification." In the last chapter of the first edition, he stated: "Species have changed, and are still slowly changing by the preservation and accumulation of successive slight favorable variations." In the second edition, Darwin modified this sentence: "Species have been modified, during a long course of descent, by the preservation or the natural selection of many successive slight favorable variations." In the fifth edition he changed this sentence by stating that modification of species occurred only "chiefly" through natural selection. In the sixth edition of *The Origin of Species*, Darwin (1872) wrote:

> *This has been effected chiefly through the natural selection of numerous successive, slight, favourable variations; aided in an important manner by the inherited effects of the use and disuse of parts; and in an unimportant manner, that is in relation to adaptive structures, whether past or present, by the direct action of external conditions, and by variations which seem to us in our ignorance to arise spontaneously.*

As Simpson (1960) noted, this is the full and final summary of Darwin's theory of evolution. It is clear that Darwin admitted four causes of evolution in sequence of importance: (1) natural selection; (2) inheritance of characters acquired by use and disuse; (3) inheritance of characters induced by direct action of the environment; (4) mutation (spontaneous variation). It should be noted that here Darwin regarded the direct action of the environment as unimportant factor. Many years later, Darwin changed his mind. In a letter to Moritz Wagner, Darwin wrote:

> *In my opinion, the greatest error which I have committed, has been not allowing sufficient weight to the direct action of the environment, i.e. food, climate, &c., independently of natural selection. When I wrote* The Origin, *and from some years afterwards, I could find little good evidence of the direct action of the environment; now there is a large body of evidence.*
>
> ***Darwin (1888)***

Unfortunately, many people failed to understand the principle of natural selection, and regarded it as the only cause of evolution. Darwin became very angry when he was accused to consider only selection as a factor in evolution. In a letter to *Nature* in responding to a critique by Sir Wyville Thomson, Darwin (1880) strongly rejected the assertion that his theory of species change depended only on natural selection:

> *Can Sir Wyville Thomson name anyone who has said that the evolution of species depends only on natural selection? As far as concerns myself, I believe that no one has brought forward so many observations on the effects of the use and disuse of parts, as I have done in my* 'Variation of Animals and Plants under Domestication' ... *I have likewise there adduced a considerable body of facts, showing the direct action of external conditions on organism.*

2.3 An Interesting Parceling Out of Darwinism

In the late 19th century, Darwin's four factors were parceled out into three distinct theories, each emphasizing one or two of those factors at the expense of the others. (1) Neo-Darwinism, which emphasized Darwin's first factor, natural selection, and strongly rejected the inheritance of acquired characters, whether acquired from use and disuse or from environmental influence. (2) Neo-Lamarckism, which emphasized the inheritance of acquired characters, Darwin's second and third factors, and minimized without necessarily wholly rejecting the influence of natural selection and of mutation. (3) Mutationism, which emphasized Darwin's fourth factor, what we now call mutation. To Darwin, mutation was only the least important way, in which materials for evolution arose. One of the best known mutationists, de Vries, maintained that mutations were the only way in which significant evolutionary change occurs. They reduced natural selection to the minor and negative role of eliminating mutants so grossly malformed as to be inviable, and agreed with the Neo-Darwinians in denying the inheritance of acquired characters (Simpson, 1960).

3. NEO-DARWINISM ACCEPTS NATURAL SELECTION BUT DENIES LAMARCKIAN INHERITANCE

3.1 Weismann and Wallace: Two Pioneers of Neo-Darwinism

Actually, the term "neo-Darwinism" was invented by Romanes (1896–1897) to distinguish between Darwin's original theory and the version of the theory that excluded the inheritance of acquired characters. According to Mayr (1985), Weismann was one of the towering figures in the history of evolutionary biology, and had the greatest impact on evolutionary theory after Darwin. It was he who was responsible for what Romanes called neo-Darwinism. He was a believer in natural selection as the one controlling process of evolution; but, unlike Darwin, he denied the inheritance of acquired characters. He claimed that natural selection was the sole mechanism of evolution. Now Weismann's theory has been seriously challenged (Eaton et al., 2015; Pollard, 1984).

Another famous neo-Darwinist was Wallace. According to Romanes (1889), from the time when Darwin and Wallace published their theory of evolution by natural selection, they did not agree upon certain points in relation to the causes of evolution. "It was the opinion of Darwin that natural selection has been the chief, but not the only, cause of organic evolution, while, in the opinion of Wallace, natural selection has been the all and in all of such evolution, virtually the sole and only principle which has been concerned in the development both of life and of mind from the ameba to the ape." It should be noted that after Weismann denied the inheritance of acquired characters, which Darwin had accepted and which his Pangenesis was constructed to explain, Wallace was compelled to discard Darwin's view in favour of that of Weismann (Wallace, 1908). The similar viewpoint was held by Galton and the research school known as biometrics (Bowler, 2005; Gayon, 1998).

Bailey (1894) compared the Darwinism and neo-Darwinism, and found one big difference between them: "It is worthy of remark that Darwinismn or Neo-Lamarckism sees first the facts or phenomena and then tries to explain them; while Neo-Darwinism or Weismannism assumes first a hypothesis and then tries to prove it. I think that any one will be struck with this difference of attitude, if he read Darwin's chapter upon Pangenesis, and then read Weismann's essay upon heredity."

In a *Science* paper, Ortmann (1910) also noticed one defect of Weismannism:

> *Weismann's opinion to the contrary and his idea of 'germinal variation' is a working hypothesis pure and simple, and should be used only as such; but the two opposite views should never be used as evidence against each other, and this is what Weismann does again and again, also in the present book. The Lamarckian principle is wrong, because it is in conflict with the Weismannian theory of the germ plasm, and the latter is correct, because, since the Lamarckian theory is wrong, it is the only way to explain evolution. This is practically the essence of Weismann's argumentation: a schoolboy's blunder against logic.*

3.2 The Modern Synthesis of Evolutionary Theory

As Futuyma (1984) pointed out, neo-Darwinism began with Weismann's rejection of the Lamarckian inheritance of acquired characters and matured in the so-called modern synthesis, which combined Darwinian natural selection, Mendelian genetics, mutation and other fields into a consensus about how evolution occurs. Many historians admit three major stages in the development of modern synthesis: (1). Darwinism in which natural selection, inheritance of acquired characters and mutation were causes of evolution; (2) neo-Darwinism, referring to Weismann's and Wallace's complete rejection of Lamarckian inheritance of acquired characters in favour of natural selection as the sole cause of evolution; and (3) the modern synthesis, which, according to Futuyma (2015), extended from about 1930 (with the publication of Fisher's *The Genetical Theory of Natural Selection*) to about 1950 (with Stebbins's *Variation and Evolution in Plants*). The most important contributors to the modern synthesis were R. A. Fisher, J.B.S. Haldane, Sewell Wright, Theodosius Dobzhansky, Julian Huxley, Ernst Mayr, George Gaylord Simpson, and G. Ledyard Stebbins (Bowler, 2005; Futuyma, 2015).

The major elements of the modern synthesis include: (1) the units of evolution are not types or single organisms, but populations of organisms; (2) evolution is based on mutations, leading to inherited variation that may be amplified by recombination; (3) natural selection, acting on inherited variation, is the major cause of evolution of adaptive characters; (4) changes in the genetic composition of populations can also result from random genetic drift, particularly in small populations; (5) new species are formed by divergence populations of an ancestral species, due to factors that reduce or prevent gene flow between populations that undergo different evolutionary changes; (6) gradual accumulation of changes by these same factors results in character differences (Futuyma, 2015). The modern synthesis, as Laland et al. (2014) pointed out, allowed the evolutionary process to be described mathematically as frequencies of genetic variants in a population change over time. That is, new variation arises through random mutation; inheritance occurs through DNA; and natural selection is the sole cause of adaptation.

3.3 Does Evolutionary Theory Need a Rethink?

Thirty years ago, Endler and McLellan (1988) proposed an approach towards a newer synthesis. According to Kutschera (2008), Carroll (2000) was the first to clearly state that data from developmental and molecular biology, geology, and the fossil record should be integrated into an "expanded evolutionary synthesis." Later, Pigliucci (2007) also emphasized that the major goal of the "expanded evolutionary synthesis" is a unification of the theory of gene and a theory of forms, which "may be achieved through an organic grafting of novel concepts onto the foundational structure of the MS, particularly evolvability, phenotypic plasticity, epigenetic inheritance, complexity theory, and the theory of evolution in highly dimensional adaptive landscapes." In addition, many other researchers also proposed that there is a need to rethink heredity and the modern synthesis (Bonduriansky, 2012; Jablonka, 2017; Koonin, 2009).

Recently, Laland et al. (2014) insisted that evolutionary theory is in urgent need of a rethink. They emphasized that "organisms are constructed in development, not simply 'programmed' to develop by genes. Living things do not evolve to fit into pre-existing environments, but co-construct and coevolve with their environments, in the process changing the structure of ecosystems." They argued that the "expanded evolutionary synthesis" could shed new light on how evolution works. The evidence supporting their position comes from allied disciplines, in particular developmental biology, genomics, epigenetics, ecology and social science. However, Wray et al. (2014) argued that "all is well" with current evolutionary theory, and that the modern synthesis remains an adequate basis for evolutionary studies. They maintained that developmental bias helps to account for how organisms adapt to their environments and diversify into many different species. They argued that none of the phenomena championed by Laland et al. (2014) were neglected in evolutionary biology. More recently, Welch (2017) proposed two possible explanations for why there have been periodic claims that evolutionary biology needs urgent reform. But Tanghe, Tiege, Pauwels, Blancke and Braeckman (2018) argued that the modern synthesis only needs to be amended, rather than replaced. Unfortunately, both sides only discussed Darwin's natural selection, but totally ignored Darwin's Pangenesis. As Kutschera (2008) pointed out, many researchers attempted to incorporate the neglected concept of symbiogenesis into the "expanded evolutionary synthesis." However, we must know that Darwin's Pangenesis is a direct precursor of symbiogenesis (Fet, 2011; Kozo-Polyansky, 2010).

In my opinion, the modern synthesis needs a rethink based on the followings points: (1) Modern synthesis is based on neo-Darwinism, and one of its cornerstones is Weismann's Germ-plasm theory, which totally rules out the inheritance of acquired characters. Now there is increasing evidence for the inheritance of acquired characters. As Noble (2015) pointed out, the central tenets of neo-Darwinism and modern synthesis have been broken. (2) According to modern synthesis, evolution is based on mutations. However, most mutations are random, homozygous recessive and harmful. Furthermore, the mutation-induced change is within the species and only forms variety, instead of species (Ahad, 2011). For example, Ortmann (1906, 1907) objected to de Vries's conception of mutation and elementary species. He pointed out that there were, indeed, forms in nature which had a tendency to breed true, but they were not isolated from other forms and should not be called species. They had been called by Darwin "varieties". (3) Modern synthesis is a graft of Darwin's theory of evolution by natural selection onto Mendelian genetics. It totally ignores Darwinian genetics, particularly Darwin's theory of heredity — Pangenesis. Now there is striking evidence in favour of Darwin's Pangenesis.

4. LYSENKOISM ACCEPTS LAMARCKIAN INHERITANCE BUT DENIES MALTHUSISM AND MUTATIONISM

4.1 Lysenko's Support of Lamarckian Inheritance and Graft Hybridization

Throughout his career, Lysenko can be regarded as a Darwinist because most of his ideas were consistent with Darwin's ideas. Lysenko recognized that Darwin was a great naturalist, the founder of scientific biology. He believed that only on the basis of Darwinism could the science of the life of plants and animals develop successfully. Lysenko's practical and experimental work on vernalization, and his theory of phasic development of plants, has been considered as genuine scientific contributions and taken seriously by scientific specialists (Roll-Hansen, 1985, 2017).

In genetics, like Darwin, Lysenko was a Lamarckist and a keen supporter of the inheritance of acquired characters (Liu, 2004). He defined heredity as "the property of a living body to require definite conditions for its life, its development and to react definitely to various conditions" (Lysenko, 1946). He held that the Lamarckian propositions were correct and scientific. At the same time he insisted that his concept on environmentally induced changes was not Lamarckism, but "Soviet Creative Darwinism" (Lysenko, 1954, pp. 589—598). According to Lysenko, the life conditions played the leading role in the development of organic world. The heredity of organisms was built up from the conditions of life over many generations. The changes of conditions of life resulted in the alterations of heredity, a process he called the "assimilations of external conditions." Once assimilated, these conditions become internal, that is, a part of heredity of the organism. He claimed to have converted spring wheat into winter wheat and *vice versa* in two to four years of autumn or spring planting, suggesting that these environmentally induced changes were transmitted to the offspring and thus led to better-adapted plant strains (Lysenko, 1946).

Lysenko greatly appreciated Darwin's notion of graft hybridization as well as Michurin's 'mentor-grafting" method. He not only recognized the existence of graft hybrid, but also applied graft hybridization to the practice of plant breeding. According to Lysenko (1954, pp. 589—598), the heredity of plants could be altered by three means: (1) graft hybridization; (2) bring external conditions to bear upon them at definite stage of development; (3) cross-breeding. Lysenko maintained that "any character may be transmitted form one strain to another by means of grafting as well as by sexual method." In grafting, according to him, the scion as well as the stock can transmit to one another hereditary characters by means of plastic substances. As he wrote, "the plastic substances produced by the scion and the stock possess the characters of the breed, are endowed with definite heredity just as chromosomes, and just as any particles of the living body." He argued that the experiments in graft hybridization "provide unmistakable proof that any particle of a living body, even the juices exchanged between scion and stock, possesses hereditary qualities." Unfortunately, he was not aware or ignored Darwin's pangenetic explanation.

Darwin (1868) held that the ill effects of inbreeding both in animals and plants could be partly or wholly removed by raising them in different environments. Lysenko's renovation of wheat seed by intravarietal crossing was based on the assumption that other plants of the same self-fertilizing strain would show slightly different environmental conditions, and that crossing them should lead to an increase in vigour. This assumption was brilliantly justified in practice, and within a few years, extensive comparative trials on many collective farms demonstrated the effectiveness of the technique in raising yields (Morton, 1951).

4.2 Lysenko's Rejection of Malthusism and Mutationism

It is well known that Darwin was partly inspired by the social economic theory of Thomas Malthus, who proposed that population growth would outstrip resource growth, thus overpopulation and insufficient supply are inevitable. Darwin applied this principle in nature: more offspring are produced than can survive; some are better suited to the prevailing conditions than others; and those better-suited individuals are more likely to leave their advantageous heritable features to the offspring (Padian, 2008). However, Lysenko challenged Darwin at this point, which he considered as erroneous aspects of Darwinism. In 1948, Lysenko gave his opening address in the meeting of the Lenin Academy of Agricultural Sciences. His first target was Malthus:

> *Many are still not clear about Darwin's error in transferring into his teaching Malthus' preposterous reactionary ideas on population .... Darwin himself was unable to fight free of the theoretical errors of which he was guilty. Today there is absolutely no justification for accepting the erroneous aspects of the Darwinian theory, those based on Malthus' theory of overpopulation with the inference of a struggle going on within species. And it is all the more inadmissible to represent those erroneous aspects as the cornerstone of Darwinism.*
>
> ***Lysenko (1954)***

Inspired by Malthus, Darwin accounted for the gap between species by intraspecific competition. But Lysenko argued that there existed no intraspecific competition but mutual assistance among individuals within a species, although he admitted that there existed interspecific competition and also mutual assistance between different species. Lysenko correctly stated how rarely intraspecific competition occurs, but he stressed this case too strongly. He ignored a very important kind of conflict between members of the same species, to which Darwin drew attention. This is the struggle, often with horns and teeth, which takes place between males of polygamous species for the possession of females. In such species the males are larger than the females. Darwin explained such facts by sexual selection (Haldane, 1948a). Evidently, Lysenko's claim of nonexistence of intraspecific competition in nature is too one-sided. Being a famous biologist, Lysenko should have been proficient in Darwin's works, but he was not, as he himself said: "[Prezent] showed me that the roots of the work I am doing lie in Darwin. And I, comrades, must confess here straightforwardly in the presence of Joseph Vissarionovich [Stalin] that of my shame I have not studied Darwin properly." (Jokes, 1995).

It is well known that Lysenko did not recognize the significance of mutation in genetics. In *The Science of Biology Today*, he wrote: "We do not deny the action of substances which produce mutations. But we insist that such action, which penetrates the organism not in the course of its development, not through the process of assimilation and disassimilation, can only rarely and only fortuitously lead to results useful for agriculture. It is not the road of systematic selection, nor the road of progressive science."

(Lysenko, 1954, pp. 589—598) Although Lysenko did not deny that chemicals could induce mutations, he felt that these could not be produced at will because they were caused by chance and not by law and therefore were not the proper base for genetics. He was suspicious of colchicine, and was reported to have said: "By treating plants with a very powerful poison, colchicine, and other torturing applications they mutilate plants." Despite Lysenko's suspicion of chemical means of inducing mutation, a few decades later, in early 1960s, Lysenko referred to radiation mutagenesis as an achievement for Soviet science (Campos, 2017).

4.3 Was Lysenko Partly Right?

It is well known that Darwin's theory of evolution by natural selection proceeds from the recognition of quantitative changes only, thus there should be no natural border lines, no discontinuity between species in nature. Lysenko opposed Darwin in this point:

> *No continuous, unbroken series of forms between species — different qualitatively definite states of living matter — have been found. This is so not because the intermediate forms in a continuous range have died out as a result of mutual competition, but because there is no such continuity in nature, nor can there be. A species is a distinct, qualitatively definite state of living matter. We must realize that speciation is a transition - in the course of the historical process — from quantitative to qualitative variations. Such a leap is prepared by the vital activity of organic forms themselves, as the result of quantitative accumulation of response to the action of definite conditions of life, and that is something that can definitely be studied and directed. The conversion of one species into another takes place by a leap.*
>
> ***Lysenko (1954)***

Lysenko's assumption was based on the following facts: In 1948, Karapetian observed in his experiments that if 28-chromosome durum wheat (*Triticum durum*) was sown late in the autumn, some of the plants are converted rather quickly, in two or three generations, into another species, into 42-chromosome soft wheat (*T. vulgare*). Moreover, several Soviet researchers found single grains of rye in durum-wheat and soft-wheat spikes (Lysenko, 1954, pp. 589—598). Lysenko's idea of sudden jumps in evolution had once supported only by a few scientists, such as Imanishi (Halstead, 1987). But recently, many people have come to realize that evolution can happen more quickly than previously thought, and can be observed within a few generations of a given species (Lallensack, 2018).

Haldane, one of most important figures of modern synthesis, made some objective comments on Lysenko. He believed that Lysenko was a very fine biologist and some of his ideas were right, though he also realized that some of Lysenko's ideas were wrong and badly wrong (Haldane, 1964). He believed that Lysenko's work on the converted wheat was a "very revolutionary discovery if true" (Haldane, 1948b). He also agreed that "one can sometimes change the character of living creature by grafting it, and the changed character is sometimes inherited" (Haldane, 1948b). Haldane (1948a) expressed partial sympathy with Lysenko over the significance of intraspecific competition. Like Lysenko, Haldane (1948a) also rejected the notion that overpopulation was the rule in nature. He never denied its existence, but maintained its extent generally exaggerated. Animal populations, Haldane (1948a) argued, were typically kept down, not by resources in short supply but by conflict with other species, suggesting that direct struggle among members of the same species is rare. Haldane argued that, from a practical standpoint, Lysenkoism was a great success. He assumed, probably correctly, that communists and non-communists alike would find it hard to believe that the Soviet regime would support Lysenko if his policies had been practical failures (Paul, 1983).

In 1977, the famous British cytogeneticist and evolutionist, Darlington, (1977) raised a question: was Lysenko a charlatan? In his book *The Lysenko Affair*, Joravsky (1970) depicted Lysenko as a pseudo-scientist from the beginning: "The painful truth is that Lysenko was never seriously involved in any genuine scientific problem". Roll-Hansen (1985) argued that Joravsky failed to recognize the big difference between Lysenko's credentials in plant physiology and in genetics, suggesting that Lysenko's work in genetics may be pseudo-science as a whole. Flegr (2002) held that Lysenko was partly right, arguing that in 1930s—1940s, Lysenko and his followers described some interesting facts, which could have been real and inspire modern researchers to extend our scientific knowledge. In my opinion, Lysenko's studies in vernalization, plant development, graft hybridization and the inheritance of acquired characters are of great interest and importance in plant physiology and genetics. His limitation is his ignorance of Darwin's work, particularly Darwin's Pangenesis. His mistake is mainly mixing science and politics. He considered Mendelian genetics as "bourgeois science" and "pseudo-science", and forced Soviet researchers to accept Michurin's teaching, for which he became notorious (Liu, 2004).

5. GOING BACK TO DARWINIAN SYNTHESIS: NATURAL SELECTION AND PANGENESIS

5.1 Darwinian Synthesis: Natural Selection and Pangenesis

Thirty years ago, van Balen (1988) mentioned the term "Darwinian synthesis," realizing that "Darwin's theory really needed a specific theory of heredity to back up the one important assumption by which it was distinguished from its rival evolutionary theories." As he wrote: "I am not claiming that Darwin's theory needed Mendel's theory to be completed; I am only claiming that Darwin's theory needed a specific theory like Mendel's theory." Unfortunately, he failed to mention Darwin's Pangenesis. As a matter of fact, natural selection and Pangenesis are Darwin's only two completely general theories comprehending all animals and plants (Hodge, 2010). In evolutionary biology, the modern synthesis based on neo-Darwinism which brought Darwinian selection together with Mendelian genetics is well known. Little is known about Darwin's own synthesis which combined his theory of evolution by natural selection with his theory of heredity and variation — Pangenesis. When talking about Darwinism, most people mean Darwin's theory of evolution by natural selection, totally ignoring his Pangenesis, which was very much part of his own Darwinism and includes the inheritance of acquired characters (Jablonka & Lamb, 1998).

It is the opinion of Ghiselin (1973) that Darwin may be regarded as the Newton of biology. He clearly demonstrated that all the phenomena of life could be related to one comprehensive theory. Ghiselin (1975) further commented: "The more one studies Darwin's works, the more apparent it becomes that he was trying to recast the whole of biology in evolutionary terms, all by himself. ... natural selection as well as Pangenesis was devised because Darwin habitually thought big." Today's world of very focused research is difficult terrain for broad-thinking scientists made in the mold of Darwin. 150 years after the publication of *The Variation of Animals and Plants under Domestication*, most people have not really entered the world into which Darwin led us.

Hodge (1985) once raised a good question: "How, then, if Darwin's theorizing ranges across so much of both Lyell's and Muller's ground, can he be represented as sending his 'evolutionary biology' into the world walking on only one real leg, natural selection, supported if at all by a flimsy physiological (including cytology) crutch, Pangenesis, where the other leg, genetics, should be?" It appears to me that natural selection and Pangenesis are two legs of Darwinism.

In 1997, when talking about the importance of internal factors in evolution, Cavalier-Smith (1997) made the following comment: "Evolutionists are very prone to cite, and maybe even read, *The Origin of Species*, but none who have not also read *The Variation of Animals and Plants under Domestication*, in which he discussed such internal factors at length, can truly appreciate the great scope and depth of Darwin's contributions to evolutionary biology." He further argued, Darwin's *The Variation* "is a very perceptive and positive seminal early contribution to the field of genome evolution, and a pointer to how the fields of genomic and organismic evolution can be integrated."

5.2 A Comparison of Lamarckism, Darwinism, Neo-Darwinism, Mutationism and Lysenkoism

Evolution is the process of heredity and variation, and variation is the fountainhead of evolution. Darwin agreed with Buffon that variations were produced by the direct action of environment, and with Lamarck by the use and disuse of parts. He explained the inheritance of acquired characters by his Pangenesis. Neo-Darwinism has overlooked the inheritance of acquired characters, graft hybridization and many important facts. Although Lysenkoism recognizes the inheritance of acquired characters and graft hybridization, it does not allow sufficient weight to natural selection and mutation. Obviously, both neo-Darwinism and Lysenkoism are too one-sided. Similarly, the original Darwinism was also far more comprehensive than both Lamarckism and Mutationism (see Table 1).

Table 1 A Comparison Among Lamarckism, Darwinism, Neo-Darwinism, Mutationism and Lysenkoism

	Lamarckism	Darwinism	Neo-darwinism	Mutationism	Lysenkoism
Natural selection	N/A	Accepted	Accepted	Rejected	Partly Rejected
Dominance	N/A	Accepted	Accepted	Accepted	Rejected
Mutation	N/A	Accepted	Accepted	Accepted	Rejected
Acquired inheritance	Accepted	Accepted	Rejected	Rejected	Accepted
Graft hybridization	N/A	Accepted	Rejected	Rejected	Accepted
Telegony	N/A	Accepted	Rejected	Rejected	Accepted
Pangenesis	N/A	Accepted	Rejected	Rejected	Rejected

5.3 Going Back to Darwinian Synthesis

In their article "Genetics and the causes of evolution: 150 years of progress since Darwin", Bonsall and Charlesworth (2010) wrote: "Ironically, genetics was Darwin's greatest failure. Despite devoting an enormous amount of effort to collecting data, he never arrived at a clear, unambiguous interpretation of the phenomena of inheritance and variation, although he observed Mendelian ratios in several crosses in his work on plants. In contrast to Gregor Mendel, his contemporary, he did not formulate a model of inheritance that made quantitative, testable predictions of the outcomes of controlled crosses, and so failed to grasp the significance of his ratios." Most recently, Meyer (2018) has also claimed that "Charles Darwin famously misunderstood inheritance, proposing that tiny particles shed from an organism's many cells migrated to the gonads, where they were transferred to offspring." These statements are obviously false and misleading, and Darwin's idea on inheritance is still neglected and misunderstood.

The modern synthesis which evolved from neo-Darwinism has been the dominant type of evolutionary thought. It accepts Mendelian genetics but totally ignores Darwinian genetics, particularly Darwin's Pangenesis. Although Mendel's laws explains dominance and has led to predictions of some new findings, it was not obvious that it also could provide a basis for the inheritance of acquired characters, graft hybridization, reversion, regeneration, xenia, telegony, inherited diseases and many other genetic phenomena. By the time Mendel's paper published, Darwin had developed his Pangenesis, a highly comprehensive theory of heredity that he constructed to account for a far wider range of phenomena than Mendel had thought. Now the discoveries of circulating cell-free DNA, mobile RNAs (particularly small RNAs), prions and extracellular vesicles provide striking evidence for Darwin's proposed gemmules. There is also increasing evidence for the inheritance of acquired characters, graft hybridization, xenia and many others facts that Pangenesis was designed to explain. Thus the neo-Darwinists were not wise to graft Darwinian evolution onto Mendelian genetics. Now neo-Darwinism and the modern synthesis are facing serious challenges (Ahad, 2011; Laland et al., 2014; Noble, 2015; Steele & Lloyd, 2015; Walsh and Huneman, 2017).

It is well known that the rediscovery of Mendel's work was one of the main theoretical contributions to the modern synthesis of evolutionary theory. In other words, the modern synthesis represents a marriage of the 19th century concept of evolution with so-called Mendelian genetics

(Pennisi, 2008). Now the rediscovery of Darwin's Pangenesis will lead our attention back to Darwin's own synthesis: natural selection and Pangenesis. It has always appeared to me that Darwinism throws more light on genetics and evolution than any other theories, and there is a need to go back to Darwinian Synthesis. This is not a historical retrogression, but an exploitation of a gold mine buried deep underground. As McLaren (1999) wrote: "History may be circular, but the history of science is helical: it repeats itself, but each time at a deeper level."

60 years ago, on the centenary of the first presentation of Darwin theory of evolution by natural selection, Michie (1958) claimed that "the third stage in genetics has begun." According to him, genetics had gone through two stages: (1) the period of classical Mendelian genetics, accompanied by the formulation of Weismann's germplasm theory and Morgan's chromosomal theory, which ignored asexual reproduction and the inheritance of acquired characters, and only had a one-way view; (2) the period of post-war genetic research, in which the inheritance of cytoplasmic factors were brought into genetics. In Michie's opinion, the third stage was just dawning in 1958. The evidence he cited included graft-induced heritable changes reported by Daniel (1929), Hall (1954), and particularly Glavinic (1955). All these reports supported Darwin's concept of graft hybridization, the striking evidence in favour of his Pangenesis. He also cited the report by Benoit, Leroy, Vendrely and Vendrely (1957), who succeeded in changing the characters of ducks of one breed by injections of DNA from another, and that the modifications continued to be identifiable in the offspring of the treated birds. It should be noted that it was the same research group who also succeeded in inducing hereditary changes in the domestic birds by blood transfusion (Leroy, Vendrely, Benoit & Vendrely, 1966), in which Galton (1871) failed to find such effects in his experiment on rabbits and thus concluded that Darwin's Pangenesis was incorrect. Michie (1958) argued that Darwin's ideas on genetics were "one hundred and ten years ahead of their time," and suggested that the day of the return of Pangenesis "has dawned." Today, 150 years after the publication of Darwin's Pangenesis, I would like to follow up with Michie (1958), and suggest that the day of her return has indeed dawned.

REFERENCES

Ahad, M. A. (2011). Molecular evolution of new species without modern synthetic theory (neo-Darwinism). *International Journal of Bio-resource and Stress Management, 2*, 131–136.

Bailey, L. H. (1894). Neo-Lamarckism and neo-Darwinism. *The American Naturalist, 28*, 661–678.

Balter, M. (2000). Was Lamarck just a little bit right? *Science, 288*, 38.

Benoit, J., Leroy, P., Vendrely, C., & Vendrely, R. (1957). Des mutations somatiques dirigées sont-elles possibles chez les Oiseaux. *Comptes rendus de l'Académie des Sciences, 244*, 2320–2321.

Bonduriansky, R. (2012). Rethinking heredity again. *Trends in Ecology & Evolution, 27*, 330–336.

Bonsall, M., & Charlesworth, B. (2010). Genetics and the causes of evolution: 150 years of progress since Darwin. *Philosophical Transactions of the Royal Society B: Biological Sciences, 365*, 2427–2429.

Bowler, P. J. (2005). Variation from Darwin to the modern synthesis. In B. Hallgrímsson, & B. K. Hall (Eds.), *Variation* (pp. 9–27). New York: Elsevier Academic Press.

Burkhardt, R. W. (2013). Lamarck, evolution, and the inheritance of acquired characters. *Genetics, 194*, 793–805.

Campos, L. (2017). Dialectics denied: Muller, Lysenkoism, and the fate of chromosomal mutation. In *The Lysenko controversy as a global phenomenon, volume 2* (pp. 164–184).

Carroll, R. L. (2000). Towards a new evolutionary synthesis. *Trends in Ecology & Evolution, 15*, 27–32.

Cavalier-Smith, T. (1997). Cell and genome coevolution: facultative anaerobiosis, glycosomesand kinetoplastan RNA editing. *Trends in Genetics, 13*, 6–9.

Cochrane, E. (1997). Viva Lamarck: A brief history of the inheritance of acquired characteristics. *Aeon, 2*, 5–39.

Daniel, L. (1929). The inheritance of acquired characters in grafted plants. In B. M. Duggar (Ed.), *Proceedings of the International Congress of Plant Sciences* (pp. 1024–1044). Menasha, Wisconsin: George Banta Publishing Company.

Darlington, C. D. (1977). Obituary – T.D. Lysenko. *Nature, 266*, 287–28.

Darwin, C. (1868). *The Variation of Animals and Plants under Domestication*. London: John Murray.

Darwin, C. (1872). *On the origin of species by means of natural selection; or the preservation of favoured races in the struggle for life* (6th ed.). London: John Murray.

Darwin, C. (1874). *The descent of man, and selection in relation to sex* (2nd ed.). London: John Murray.

Darwin, C. (1880). Sir Wyville Thomson and nature selection. *Nature, 23*, 32.

Darwin, F. (Ed.). (1888). *The life and letters of Charles Darwin*. London: John Murray.

Dodson, E. O. (1961). Neo-Lamarckism, modern Darwinism, and the origin of the vertebrates. *Journal of Paleontology, 35*, 1065–1076.

Eaton, S. A., Jayasooriah, N., Buckland, M. E., Martin, D. I. K., Cropley, J. E., & Suter, C. M. (2015). Roll over Weismann: Extracellular vesicles in the transgenerational transmission of environmental effects. *Epigenomics, 7*, 1165–1171.

Endler, J. A., & McLellan, T. (1988). The processes of evolution: Towards a newer synthesis. *The Annual Review of Ecology, Evolution, and Systematics, 19*, 395–421.

Fet, V. (2011). Symbiogenesis in Russia. In L. Margulis, C. A. Asikainen, & W. Krumbein (Eds.), *Chimeras and Consciousness: Evolution of the sensory self* (pp. 153–158). Cambridge: MIT Press.

Fine, P. E. M. (1979). Lamarckian ironies in contemporary biology. *Lancet, 313*, 1181–1182.

Flegr, J. (2002). Was Lysenko (partly) Right? Michurinist Biology in the View of Modern Plant Physiology and Genetics. *Rivista di Biologia/Biology Forum, 95*, 259–272.

Futuyma, D. J. (1984). Neo-darwinism in disfavour. *Science, 226*, 532–553.

Futuyma, D. J. (2015). Can modern evolutionary theory explain macroevolution? In E. Serrelli, & N. Gontier (Eds.), *Macroevolution: Explanation, interpretation and evidence* (pp. 29–86). Switzerland: Springer International Publishing.

Gadjev, I. (2015). Nature and nurture: Lamarck's legacy. *Biological Journal of the Linnnean Society, 114*, 242–247.

Gayon, J. (1998). *Darwinism's struggle for Survival: Heredity and the hypothesis of natural selection*. Cambridge: Cambridge University Press.

Gerould, J. H. (1922). Lamarck, Mirbel and the cell theory. *Science, 55*, 421–422.
Ghiselin, M. T. (1973). Darwin and evolutionary psychology. *Science, 179*, 964–968.
Ghiselin, M. T. (1975). The rationale of pangenesis,. *Genetics, 79*, 47–57.
Glavinic, R. (1955). *Vegetative hybridization of tomato. Proceedings of the 14th Internatinal Horticultural Congress Volume1* (pp. 440–445). Wageningen, Netherlands: H. Veenman & Zonen.
Glick, T. F., & Kohn, D. (1996). *Charles Darwin on Evolution: The development of the theory of natural selection.* Indianapolis: Hackett Publishing Company.
Haldane, J. B. S. (December 10, 1964). J. B. S. Haldane's self-obituary. *The Listener*, 934–935.
Haldane, J. B. S. (November 1, 1948a). Lysenko and Darwin. *Daily Worker*, 2.
Haldane, J. B. S. (December 9, 1948b). The Lysenko controversy. *The Listener*, 873–875.
Halstead, B. (1987). Imanishi's influence on evolution theory in Japan. *Nature, 326*, 21.
Hall, O. L. (1954). Hybridization of wheat and rye after embryo transplantation. *Hereditas, 40*, 453–458.
Hodge, M. J. S. (1985). Darwin as a lifelong generation theorist. In D. Kohn (Ed.), *The Darwinian heritage* (pp. 207–243). Princeton: Princeton University Press.
Hodge, J. (2010). The Darwin of pangenesis. *Comptes Rendus Biologies, 333*, 129–133.
Jablonka, E. (2017). The evolutionary implications of epigenetic inheritance. *Interface Focus, 7*, 20160135.
Jablonka, E., & Lamb, M. J. (1995). *Epigenetic Inheritance and Evolution: The Lamarckian Dimension.* Oxford: Oxford University Press.
Jablonka, E., & Lamb, M. J. (1998). Genic Neo-Darwinism – is it the whole story? *Journal of Evolutionary Biology, 11*, 243–260.
Jokes, T. H. (1995). Darwinist Lysenko? *Nature, 373*, 554.
Joravsky, D. (1970). *The Lysenko Affair.* Cambridge: Harvard University Press.
Koonin, E. V. (2009). The *origin* at 150: Is a new evolutionary synthesis in sight? *Trends in Genetics, 25*, 473–475.
Kozo-Polyansky, B. M. (2010). *Symbiogenesis: A new principle of evolution.* In V. Fet, & L. Margulis (Eds.). Cambridge: Harvard University Press (from the Russian edition, 1924).
Kutschera, U. (2008). From Darwinism to evolutionary biology. *Science, 321*, 1157–1158.
Laland, K., Uller, T., Feldman, M., Sterelny, K., Müller, G. B., Moczek, A., … Odling-Smee, J. (2014). Does evolutionary theory need a rethink? yes, urgently. *Nature, 514*, 161–164.
Lallensack, R. (2018). How wrap-speed evolution is transforming ecology. *Nature, 554*, 19–21.
Lamarck, J. B. (1809). *La Phlilosophie zoologique* (translated by Hug Elliot, 1914, as *Zooligical Philosophy*. London: Macmillan). New York: Stechert-Hafner. Reprint 1963.
Leroy, P., Vendrely, R., Benoit, J., & Vendrely, C. (1966). Divergences observed in the descendants of Rhode Island Red chickens M-44 after injections of specifically different bloods or fractions of blood. In E. A. Duyunov, & G. Kopylovskaya (Eds.), *The Thirteenth World's Poultry Congress Proceedings* (pp. 106–109). Kiev: USSR.
Liu, Y.-S. (2004). Lysenko's contributions to biology and his tragedies. *Rivista di Biologia/ Biology Forum, 97*, 483–498.
Lysenko, T. D. (1946). *Heredity and its variability, trans. Theodosius Dobzhansky.* New York: King's Crown Press.
Lysenko, T. D. (1954). *Agrobiology.* Moscow: Foreign Languages Publishing House.
Mayr, E. (1985). Weismann and evolution. *Journal of the History of Biology, 18*, 295–329.
McLaren, A. (1999). Too late for the midwife toad: Stresss, variability and Hsp90. *Trends in Genetics, 15*, 169–171.
Meyer, A. (2018). Our inheritance. *Science, 360*, 863.
Michie, D. (1958). The third stage in genetics. In S. S. Barnett (Ed.), *A Century of Darwin* (pp. 56–84). London: Heinemann.
Morton, A. G. (1951). *Soviet genetics.* London: Lawrence & Wishart.

Noble, D. (2015). Central tenets of neo-Darwinism broken. Response to 'neo-Darwinism is just fine'. *Journal of Experimental Biology, 218*, 2659.
Ortmann, A. E. (1906). The fallacy of the mutation theory. *Science, 23*, 746–748.
Ortmann, A. E. (1907). Facts and interpretations in the mutation theory. *Science, 25*, 185–190.
Ortmann, A. E. (1910). Weismannism, a criticism of die selektions-theorie. *Science, 31*, 815–819.
Padian, K. (2008). Darwin's enduring legacy. *Nature, 451*, 632–634.
Paul, D. B. (1983). A war on two fronts: J. B. S. Haldane and the response to lysenkoism in britain. *Journal of the History of Biology, 16*, 1–37.
Pennisi, E. (2008). Modernizing the modern synthesis. *Science, 321*, 196–197.
Pigliucci, M. (2007). Do we need an extended evolutionary synthesis? *Evolution; International Journal of Organic Evolution, 61*, 2743–2749.
Pollard, J. W. (1984). Is Weissmann's barrier absolute? In M.-W. Ho, & P. T. Saunders (Eds.), *Beyond Neo-Darwinism. An introduction to the new evolutionary paradigm* (pp. 291–314). London: Academic Press.
Por, F. D. (2006). The actuality of Lamarck: Towards the bicentenary of his *Philosophie zoologique*. *Integrative Zoology, 1*, 48–52.
Roll-Hansen, N. (1985). A new perspective on Lysenko? *Annals of Science, 42*, 261–278.
Roll-Hansen, N. (2017). Organicism and lysenkoism. *Studies in History and Philosophy of Biological and Biomedical Sciences, 65*, 30–34.
Romanes, G. J. (1889). Mr. Wallace on darwinism. *Science, 14*, 150–155.
Romanes, G. J. (1896-1897). *Darwin and after Darwin: An exposition of the Darwinian theory and a discussion of post-Darwinian questions*. Chicago: Open Court Publishing Company.
Simpson, G. G. (1960). The world into which Darwin led us. *Science, 1960*(181), 966–974.
Stafleu, F. A. (1971). Lamarck: the birth of biology. *Taxon, 20*, 397–442.
Steele, E. J., Lindley, R. A., & Blanden, R. V. (1998). *Lamarck's Signature: How retrogenes are changing Darwin's natural selection paradigm*. Massachusetts: Perseus Books.
Steele, E. J., & Lloyd, S. S. (2015). Soma-to-germline feedback is implied by the extreme polymorphism at IGHV relative to MHC. *BioEssays: News and Reviews in Molecular, Cellular and Developmental Biology, 37*, 557–569.
Tanghe, K. B., Tiege, A. D., Pauwels, L., Blancke, S., & Braeckman, J. (2018). What's wrong with the modern evolutionary synthesis? A critical reply to Welch (2017). *Biology & Philosophy, 33*, 23.
van Balen, G. (1988). The darwinian synthesis: A critique of the rosenberg/Williams argument. *The British Journal for the Philosophy of Science, 394*, 441–448.
Wallace, A. R. (1908). *My life: A record of events and opinions*. London: Chapman & Hall.
Walsh, D. M., & Huneman, P. (2017). Introduction: Challenging the modern synthesis. In D. M. Walsh, & P. Huneman (Eds.), *Challenging the modern Synthesis: Adaptation, development, and inheritance* (pp. 1–39). Oxford Scholarship Online.
Welch, J. J. (2017). What's wrong with evolutionary biology? *Biology & Philosophy, 32*, 263–279.
Wray, G. A., Hoekstra, H. E., Futuyama, D. J., Lenski, R. E., Mackay, T. F., Schluter, D., & Strassmann, J. E. (2014). Does evolutionary theory need a rethink? No, all is well. *Nature, 514*, 161–164.

FSC
www.fsc.org
MIX
Papier | Fördert
gute Waldnutzung
FSC® C083411

Kolibri 360 GmbH
Ferdinand-Jühlke-Straße 7
99095 Erfurt, Deutschland
produktsicherheit@kolibri360.de